coil wound

THE FREE-ENERGY DEVICE HANDBOOK

A Compilation of Patents & Reports

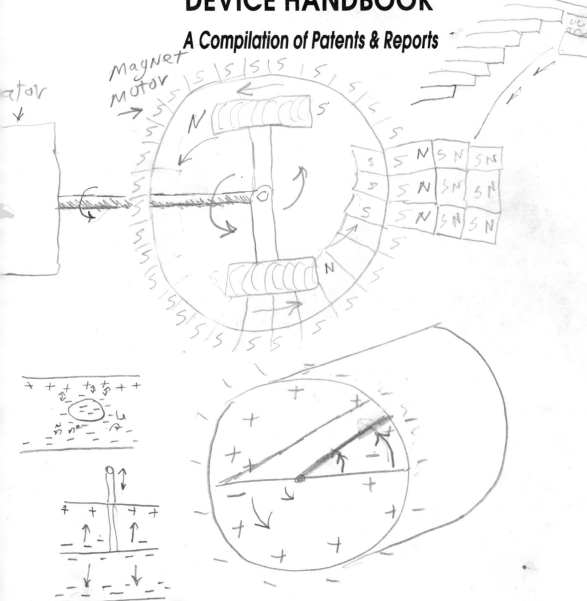

Magnet Motor

ator

Altrenate charges
to produce movement

THE FREE-ENERGY DEVICE HANDBOOK

A Compilation of Patents & Reports

compiled by David Hatcher Childress

Adventures Unlimited Press

This book is dedicated to the scientists and engineers who
continue to forge ahead despite opposition from all sides.

Special thanks to Carl Hart & North Light Press, Jackie
Munguia, Kathy Collins, Duncan Roads, and many more!

The Lost Science Series:
 •THE ANTI-GRAVITY HANDBOOK
 •ANTI-GRAVITY & THE WORLD GRID
 •ANTI-GRAVITY & THE UNIFIED FIELD
 •EXTRATERRESTRIAL ARCHAEOLOGY
 •THE FANTASTIC INVENTIONS OF NIKOLA TESLA
 •VIMANA AIRCRAFT OF ANCIENT INDIA & ATLANTIS
 •MAN-MADE UFOS: 1944-1994
 •TAPPING THE ZERO POINT ENERGY
 •THE ENERGY GRID
 •THE BRIDGE TO INFINITY
 •ETHER TECHNOLOGY

TABLE OF CONTENTS

1

Nikola Tesla and Free-Energy

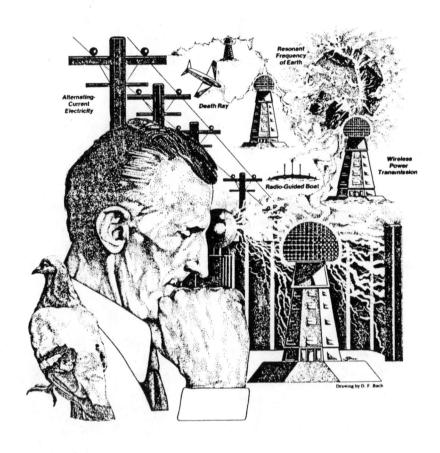

Alternating-
Current
Electricity

Death Ray

Resonant
Frequency
of Earth

Wireless
Power
Transmission

Radio-Guided Boat

Drawing by D. F. Bach

Nikola Tesla's "Free Energy" Documents

Oliver Nichelson
333 North 760 East
American Fork, Utah 84003

Ten years after patenting a successful method for producing alternating current, Nikola Tesla claimed the invention of an electrical generator that would not "consume any fuel." Such a generator would not have an external prime mover such as steam or falling water.

The documents that establish Tesla's involvement in this line of research are presented below.

On June 9th, 1902, the *New York Times* and the *New York Herald* carried a story of a Clemente Figueras, a "woods and forest engineer," who had invented a device for generating electricity without burning any fuel. In the Nikola Tesla Collection, at Columbia University Library is a letter from the inventor to his friend Robert Underwood Johnson, the editor of *Century* Magazine, with a clipping of the *Herald* article enclosed.[1]

In the three page letter Tesla states that he suggested such a generator in his *Century* magazine article, and that he has worked on such a design for sometime (Figures 2 & 3).

1. *N.Y. Herald,* June 9, 1902

Page 200 from the June 1900 *Century* article is reproduced below. Tesla once called this article the most important that he wrote. The "novel facts" citation mentioned in the letter is found in the first column, next to the last paragraph, first sentence.[2] Discussion of the "novel facts" just precedes the article's subsection dealing with a "'Self-Acting' Machine...Capable...of Deriving Energy From the Medium."

200   THE CENTURY MAGAZINE.

tions show, with the approach to the center at the rate of approximately 1° C. for every hundred feet of depth. The difficulties of sinking shafts and placing boilers at depths of, say, twelve thousand feet, corresponding to an increase in temperature of about 120 C., are not insuperable, and we could certainly avail ourselves in this way of the internal heat of the globe. In fact, it would not be necessary to go to any depth at all in order to derive energy from the stored terrestrial heat. The superficial layers of the earth and the air strata close to the same are at a temperature sufficiently high to evaporate some extremely volatile substances, which we might use in our boilers instead of water. There is no doubt that a vessel might be propelled on the ocean by an engine driven by such a volatile fluid, no other energy being used but the heat abstracted from the water. But the amount of power which could be obtained in this manner would be, without further provision, very small.

Electricity produced by natural causes is another source of energy which might be rendered available. Lightning discharges involve great amounts of electrical energy, which we could utilize by transforming and storing it. Some years ago I made known a method of electrical transformation which renders the first part of this task easy, but the storing of the energy of lightning discharges will be difficult to accomplish. It is well known, furthermore, that electric currents circulate constantly through the earth, and that there exists between the earth and any air stratum a difference of electrical pressure, which varies in proportion to the height.

In recent experiments I have discovered two novel facts of importance in this connection. One of these facts is that an electric current is generated in a wire extending from the ground to a great height by the axial, and probably also by the translatory, movement of the earth. No appreciable current, however, will flow continuously in the wire unless the electricity is allowed to leak out into the air. Its escape is greatly facilitated by providing at the elevated end of the wire a conducting terminal of great surface, with many sharp edges or points. We are thus enabled to get a continuous supply of electrical energy by merely supporting a wire at a height, but, unfortunately, the amount of electricity which can be so obtained is small.

The second fact which I have ascertained is that the upper air strata are permanently charged with electricity opposite to that of

the earth. So, at least, I have interpreted my observations, from which it appears that the earth, with its adjacent insulating and outer conducting envelop, constitutes a highly charged electrical condenser containing, in all probability, a great amount of electrical energy which might be turned to the uses of man, if it were possible to reach with a wire to great altitudes.

It is possible, and even probable, that there will be, in time, other resources of energy opened up, of which we have no knowledge now. We may even find ways of applying forces such as magnetism or gravity for driving machinery without using any other means. Such realizations, though highly improbable, are not impossible. An example will best convey an idea of what we can hope to attain and what we can never attain. Imagine a disk of some homogeneous material turned perfectly true and arranged to turn in frictionless bearings on a horizontal shaft above the ground. This disk, being under the above conditions perfectly balanced, would rest in any position. Now, it is possible that we may learn how to make such a disk rotate continuously and perform work by the force of gravity without any further effort on our part; but it is perfectly impossible for the disk to turn and to do work without any force from the outside. If it could do so, it would be what is designated scientifically as a "perpetuum mobile," a machine creating its own motive power. To make the disk rotate by the force of gravity we have only to invent a screen against this force. By such a screen we could prevent this force from acting on one half of the disk, and the rotation of the latter would follow. At least, we cannot deny such a possibility until we know exactly the nature of the force of gravity. Suppose that this force were due to a movement comparable to that of a stream of air passing from above toward the center of the earth. The effect of such a stream upon both halves of the disk would be equal, and the latter would not rotate ordinarily; but if one half should be guarded by a plate arresting the movement, then it would turn.

A DEPARTURE FROM KNOWN METHODS—POSSIBILITY OF A "SELF-ACTING" ENGINE OR MACHINE, INANIMATE, YET CAPABLE, LIKE A LIVING BEING, OF DERIVING ENERGY FROM THE MEDIUM—THE IDEAL WAY OF OBTAINING MOTIVE POWER.

WHEN I began the investigation of the subject under consideration, and when the preceding or similar ideas presented themselves

Century, June 1900, p. 200.

A careful examination of the article reveals the inventor believed his design for an electrical generator which is its own prime mover, that is, does not "consume any fuel," would not violate the energy conservation principle. Tesla believed, rather, that his design transformed one form of energy into another.

Earth's Magnetic Field

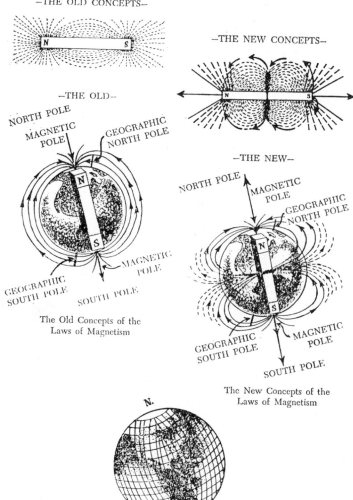

-THE OLD CONCEPTS-

-THE NEW CONCEPTS-

-THE OLD-

NORTH POLE

MAGNETIC POLE

GEOGRAPHIC NORTH POLE

GEOGRAPHIC SOUTH POLE

MAGNETIC POLE

SOUTH POLE

The Old Concepts of the Laws of Magnetism

-THE NEW-

NORTH POLE

MAGNETIC POLE

GEOGRAPHIC NORTH POLE

GEOGRAPHIC SOUTH POLE

MAGNETIC POLE

SOUTH POLE

The New Concepts of the Laws of Magnetism

N.

S.

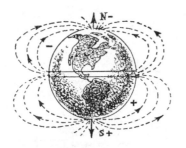 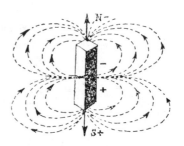

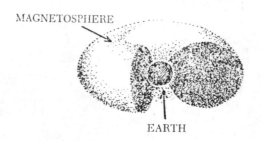

MAGNETOSPHERE

EARTH

From the book, MAGNETISM AND
ITS EFFECTS ON THE LIVING SYSTEM
by Albert Davis and Walter Rawls, Jr.
Here we see how the Earth's Magnetic
Field creates an electro-magnetic
grid around the earth. What does this
grid have to do with the earth's
gravitational field?

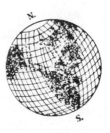

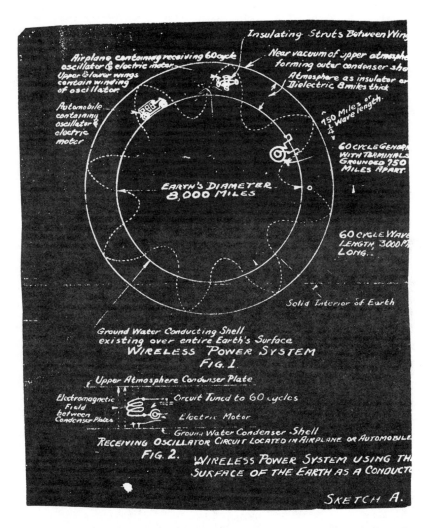

Tesla's sketch of the terrestrial resonator operating at a frequency of 60Hz. He has drawn a standing wave with 8 cycles. This gives the fundamental as 60/8 = 7.5Hz. The sketch is from a memo in 1925

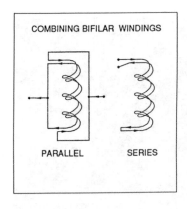

COMBINING BIFILAR WINDINGS

PARALLEL   SERIES

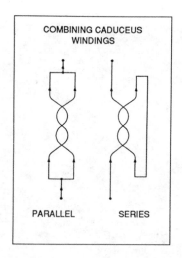

COMBINING CADUCEUS
WINDINGS

PARALLEL   SERIES

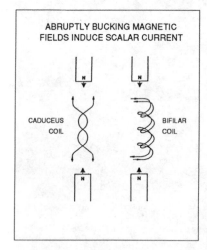

ABRUPTLY BUCKING MAGNETIC
FIELDS INDUCE SCALAR CURRENT

CADUCEUS
COIL

BIFILAR
COIL

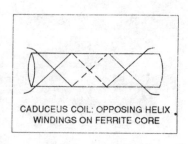

CADUCEUS COIL: OPPOSING HELIX
WINDINGS ON FERRITE CORE

Unified Field Quadrangle and Energy Gradient

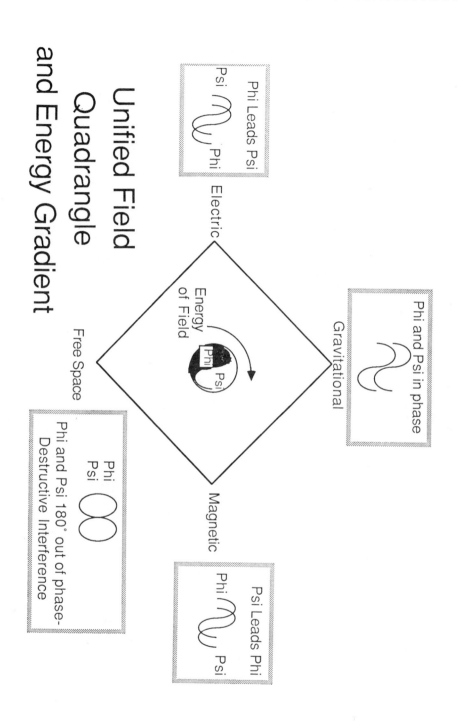

TESLA TECHNOLOGY AND RADIOISOTOPIC ENERGY GENERATION

BY
PAUL M. BROWN

JUNE 27, 1990

PAPER PRESENTED AT THE
1990 TESLA SYMPOSIUM
JULY 26-29, 1990
COLORADO SPRINGS, COLORADO

Nucell, Inc.
A Subsidiary of Peripheral Systems, Inc.

Peripheral Systems, Inc.

TESLA TECHNOLOGY AND RADIOISOTOPIC ENERGY GENERATION

BY
PAUL M. BROWN

JUNE 27, 1990

The basic electrical resonance principles pioneered by Nikola Tesla in the late 1800's are now being applied to a method of generating electricity from natural radioactive decay. Nucell, Inc. a subsidiary of Peripheral Systems, Inc.of Portland, Oregon, received a patent May, 1989, for their resonant nuclear oscillator (Figure I). In general, the resonant nuclear oscillator is an LCR tank circuit tuned to oscillate at its self resonant frequency. Energy in excess of the operational losses is contributed from a radioactive source to the tank circuit through a phenomenon known as the Beta Voltaic Effect. Net electric current is then removed from the oscillator through an impedance matched transformer to deliver high frequency electricity in usable form to drive a load.

The Beta Voltaic Effect may simply be defined as the conversion of ionizing radiation to electrical energy by a material or combination of materials. Radiation that is absorbed in the vicinity of any potential barrier will generate separated electron-hole pairs which in turn flow in an electric circuit in response to the influence of the electric potential field.

Radioactive decay energy is several orders of magnitude greater than chemical energy. For this reason, this technology has promise of yielding low volume, low weight, high energy density power sources that will be economical for long unattended life with high reliability.

Devices for converting natural radioactive decay directly into electricity are nothing new (Figure 2). The Beta Cell was first demonstrated by Moseley in 1913 and over the years many types and methods have been developed. This technology has been made possible due to the electrical nature of alpha and beta disintegrations.

(Figure 3) The simplest form of nuclear battery is the Burke Cell. This method consists of a conventional battery and a conventional load connected by means of a radioactive conductor. If we inspect this arrangement we find that all of the power dissipated in the load is not drawn from the battery. And upon closer examination we find that a current amplification occurs within the radioactive conductor.

This phenomenon is known as the Beta Voltaic Effect and may be explained by referring to Figure 4. For the simple case of this example, we will set the radioactive source (any alpha or beta emitter) external and separate from a silver wire. Now the battery from Figure 3 provides an electromotive force (emf) across the wire and consequently, conduction electrons within the wire are set in uniform motion. By definition, electricity is measured in terms of the number of charged particles (electrons) moving past a point in a unit of time and we call this amperes.

The process by which a beta particle is absorbed, is such that the beta particle collides with the molecular structure of the copper knocking electrons free. This electron avalanche occurs until the beta particle (electron) effectively comes to rest. A single beta particle emitted from strontium-90 that is absorbed in copper will generate 80,000 ions in a distance of .030 inches. Now, as soon as these electrons are knocked loose, they effectively become free electrons in the wire, and as such these additional electrons are acted upon by the emf applied across the wire to give the avalanche electrons a uniform direction of flow, regardless of their incident angle. This increase in the number of moving charge carriers is measured in the real world

as increased current. We also measure a reduction in the resistance of the wire, and an increase in its conductivity while the current is directly proportional to the voltage. In other words, the current goes up with an increase in voltage. This is basically attributed to the increased emf acting on a greater number of avalanche electrons.

(Figure 5) a cartoon representation of the basic beta voltaic converter is shown. Electrode A has a positive potential while Electrode B is negative with the potential difference provided by any conventional means. An electric field exists between the electrodes and we shall call this zone the junction. The junction between the two electrodes is thus comprised of a suitably ionizable medium exposed to decay particles emitted from a radioactive source.

In general, the introduction of ions from any source into an electric field will generate electricity in accordance with well-known physical or chemical principles and may be satisfactorily explained in terms commonly associated with the Beta Voltaic Effect. The energy contributed to such a circuit does not come from the ions themselves but rather from the work done on the circuit to generate the ions, known as the ionization potential of that particular material.

An amount of work must be performed on a neutral atom to remove electrons (ionize the atom). This work manifests itself as increased potential energy and may be utilized to do work before allowing the electron and ion to recombine.

Neither the electric field, the electrodes or the medium between the electrodes contribute any energy in the Beta Voltaic Effect. The energy is contributed by the ion generator whether this mechanism is chemical, electromagnetic or nuclear is irrelevant.

In other words, assume the conductor is irradiated with beta particles. As these particles penetrate the conductor, collisions occur with electrons in the lattice of the conductor resulting in the transfer of energy to these electrons and exciting them to a higher energy level in the conduction band.

Now we will look at how we apply this phenomenon to our device. Figure 6 depicts a basic LC tank circuit comprised of an inductor and a capacitor. Theoretically, if this LC circuit were superconductive, then an externally applied electric impulse would yield an LC oscillation that would continue to ocillate forever due to no losses in the system.

However, our LC circuit is not superconductive and the oscillation damps out due to the losses inherent to the LC tank. To minimize these inherent losses, we tune the circuit into resonance at the self-resonant frequency of the inductor.This causes the inductive and capacitive reactances to cancel leaving only ohmic losses (resistance).

(Figure 7) If we apply a radioactive source as part of the LC tank, then through every cycle of the oscillation of which current is flowing, that current gets amplified by an mount proportional to the activity of the source. All we need is an input of an amount of energy equal to the system losses to achieve a sustained oscillation. At this point, we have a self-driven oscillator what we call a Nuclear Powered Oscillator.

Any energy contributed to this oscillating LC tank must be removed and we accomplish this (Figure 8) by simply impedance matching a transformer which yields high frequency AC current to drive a load. In a nutshell, that is the principle of operation for the Resonant Nuclear Power Supply an LC tank circuit oscillating at its self-resonant frequency, driven by natural radioactive decay energy. Energy in excess of the operational requirements is removed through a transfer to yield electrical energy in usable form to drive a load.

Figure 9 depicts the starting method which involves the use of a high voltage source to charge the capacitor of the tank circuit, which is then discharged to ground through a Class C amplifier at a rate equal to the resonant frequency of the tank circuit. A spectrum analyzer is used to monitor the activity within the tank and once a clean oscillation is started, the high voltage power supply and Class C amplifier are removed; a process that takes a few seconds, then the power removed from the tank circuit is determined by measuring the voltage drop across a resistor of known value and double-checked by directly measuring the current delivered to the load.

The great attraction of radioisotope generators lies in the fact that isotope energy densities are several orders of magnitude greater than chemical energy density. However, the technology currently in use for radioisotope power generation is severely limited by its low efficiency, isotope limitations and heavy shielding requirements, while a resonant nuclear generator does not suffer from these limitations.

(Figure 10) Here we have the actual component layout of an early resonant nuclear power supply. We can see the radioactive source and its mount along with the primary inductor and matching transformer. The tuning capacitors are not shown.

(Figure 11) This is the actual wiring of the prototype shown in the previous slide. Although this generated electricity, it also demonstrated a frequency stability problem and showed signs of material degradation.

Economic studies indicate that a radioisotopic nuclear battery is economically competitive with chemical batteries for applications requiring lifetimes of over two years at remote locations where the expense of charging or changing batteries is significant. Applications where the inaccessibility after implantation is a consideration that leads to selection of nuclear batteries due to their superior reliability and life.

We have pursued several design variations and are currently working with an independent nuclear engineering firm. Our current program will generate engineering data in the coming months. Of course any alpha or beta emitting isotope will work while a design variation also allows the use of gamma sources. We have experimented with cesium, strontium, radium, krypton, tritium, promethium and probably some others. All these sources have worked, however, for personnel safety and application considerations we are currently planning to use krypton-85 as the fuel source, although strontium-90 is also a good candidate.

Large quantities of krypton-85 are contained in stored power-reactor fuels and about 1 MegaCurie per year is available from processed fuel. It is estimated that 42 MegaCuries of krypton-85 could be obtained from existing inventories in power-reactor fuels with the content in spent power-reactor fuels at about 8,500 Curies per ton.

Of the many radioactive isotopes generated by the fission of uranium, krypton-85 has many unique properties of which the most Important is its advantage of being environmentally the most acceptable radioisotope available for power production.

Preliminary data suggests that energy densities on the order of .25 watt per cubic centimeter is achievable.

Market surveys have been conducted by the nuclear industry in the past and the conclusion has been that there is a need for long-life radioisotope nuclear batteries. Of course, economic and logistical factors must be considered for comparison (Figure 12).

Obviously the physical size and shape of the radioisotope batteries will need to be related to the intended applications. For example, in oil and gas well drilling, there is an increasing benefit in continuously measuring and monitoring geo-physical data from the bottom of the hole. Under these circumstances it would be necessary

to accommodate the battery within the diameter of the drilling tubes. Another important potential market is to supply long life, electrical service to the sonar detectors which are in locations throughout the oceans of the world. The overall configuration in the case would probably be quite different. All of these possible applications need to be considered throughout the development phase.

Based on applications surveys, a design, development and testing program is being conducted on novel radioisotopic batteries that will be economical for long unattended life with high reliability, low weight and volume in the power range of 10 to 5,000 mW (e).

References:

(1) Brown, Paul. THE MORAY DEVICE AND THE HUBBARD COIL WERE NUCLEAR BATTERIES in Magnets In Your Future Magazine, Vol. 2, No. 3, March 1987.

(2) Brown, Paul. RESONANT NUCLEAR POWER SUPPLY in Raum & Zeit Magazine,Vol. 1, No. 2, August-September, 1989.

(3) Brown, Paul. American Nuclear Society 1989 Winter Meeting, San Francisco, California, November 26-30, 1989, RESONANT NUCLEAR BATTERY MAY AID IN MITIGATING THE GREENHOUSE EFFECT.

(4) Brown, Paul. American Nuclear Society Annual Meeting, June 10-14, 1990, Nashville, Tennessee, THE BETA VOLTAIC EFFECT APPLIED TO RADIOISOTOPIC POWER GENERATION.

For More Information Contact;

Paul Brown
Nucell, Inc.
12725 S.W. 66th Avenue, Suite 102
Portland, Oregon 97223
503/624-8586

NUCELL, INC. A SUBSIDIARY OF PERIPHERAL SYSTEMS, INC.

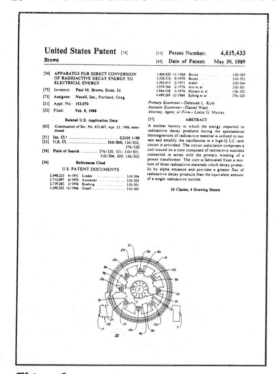

Figure 1

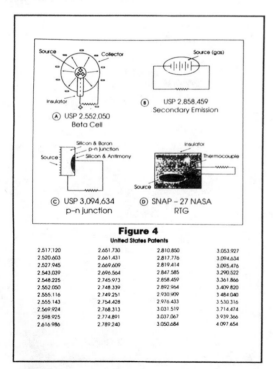

Figure 2

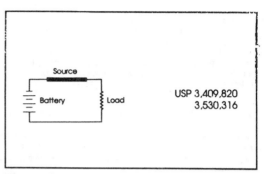

Figure 3

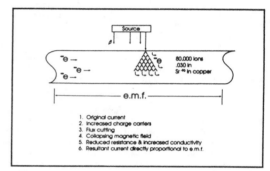

Figure 4

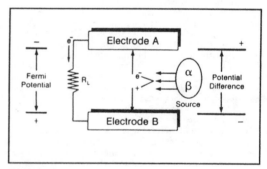

Figure 5

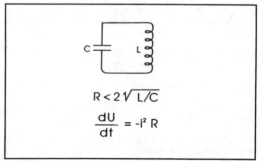

Figure 6

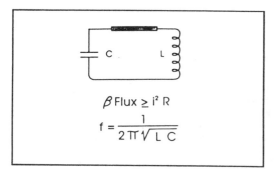

$$\beta \text{ Flux} \geq i^2 R$$

$$f = \frac{1}{2\pi\sqrt{LC}}$$

Figure 7

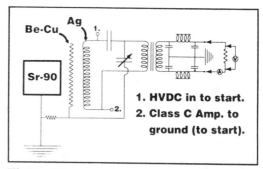

1. HVDC in to start.
2. Class C Amp. to ground (to start).

Figure 11

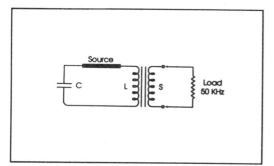

Figure 8

Figure 12

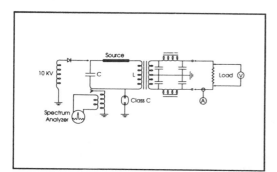

Figure 9

Figure 10

VARIABLE RELUCTANCE ALTERNATOR

John Ecklin - Flux Switch Alternator U. S. Pat. No. 4,567,407

The patent number for the bottom of the previous page is 4,567,407 and it was granted on Jan. 28, 1986. Fig. 3A and 3B depicts how the magnetic fields are reversed in both AC output coils simultaneously. The preferred embodiment for Fig. 4 is to have no windings on the rotor (5). The rotor is then made up of steel laminations. This is what in the 1890's was called a flux switch alternator. It had no brushes in this century old technology.

Patent 4,567,407 combines this old technology with the newer electronically commutated motor controllers. Sensors determine the position of the rotor and increse the saturation of the stator from 80% to 98% to pull the rotor in faster than normal. This gives a motor action and since the saturation of the stator is increased the required power is automatically captured by the AC output coils. Fig. 1 demonstrates the principle. 19 is a 3/8ths inch diameter ball bearing on top of a ½" diameter by ¼" thick ceramic button magnet resting on a horizontal steel surface 23. When the ball is pulled to the edge of the magnet and is released you will see a highly damped oscillatory motion. Turn the magnet over and you see the same thing. The ball is equivalent to either rotor pole and the magnet is equivalent to any stator pole. In other words the patent pulls the rotor to the stator and you no longer have to use input torque to force the rotor to the stator. This is how you skirt Lenz's Law compared to all of today's Faraday generators.

If you tie a very springly (high steel content) paper clip to a 6" thread you can actually see the source of the energy which is unpaired electron spin in iron atoms. With practice clip can lift ball from magnet in less than 1/10th of a second and will hang there for 50 years and more. How can we store enough energy in the clip in 1/10th second to keep the ball from falling for 50 years? We can't. The energy is already in the iron atoms of the crystal or clip. The magnet merely sets the direction of spin of most of the 4 unpaired electrons in most of the atoms in the clip. As long as the clip and ball stay together these electrons keep spinning in the same direction. If you ever separate the clip and ball you will have to use the magnet again before clip will lift ball.

Since all electrons in all atoms spin on their axis with the same angular momentum each one is an infinite source of energy. I call this God's perfect flywheel. Something about his atoms always keep the electrons spinning at the same rate. This patent is an over-unity device from the standpoint of torque but it is way, way under unity when we consider the energy in electron spin. This atomic energy as we do not change the atom eternally by splitting or joining atoms as in fission and fusion which are nuclear energy and very polluting.

b) Paul Brown, Bliss, Idaho   (June, 1982)

Paul Brown, as an independent researcher, has accomplished some significant project work in the area of John Ecklin's original S.A.G. concept, by expanding on the basic principles involved in the functioning of the S.A.G.'s.

His *Magnetic distributor Generator,* which is also known as a *Variable Reluctance Alternator,* consists of utilizing both D.C. input coils and A.C. output coils wound on 90 degree crossed laminations. The iron laminations are in the form of an exact ninety degree cross-over so that exactly opposite North-South magnetic poles are established, as in normal two-pole D.C. motor design.

A split iron and aluminum rotor provides the alternating make and break magnetic circuits between the D.C. and A.C. iron laminations, and their corresponding D.C. and A.C. coils, respectively. When the matching arms of the rotor close the gap between the stator laminations, a magnetic flux flows through the closed circuit, which causes E.M.F. flow within one set of (A.C.) of coils, which are opposite to each other.

As the rotor is turned through ninety degrees, this magnetic gap opened and the magnetic flux and corresponding E.M.F. in the coil cease. Since the iron laminations with their D.C. coils are energized by a D.C. input, this D.C. input is converted (through basic transformer action) to an A.C. flow by the uniform making and breaking of the A.C. iron laminations and their associated A.C. coils.

The characteristics of the Variable Reluctance Alternator are as follows:

1) Voltage increase with an increase in R.P.M.

2) Voltage increases with the number of turns of wire on the output coil (per transformer theory). (The project work of R. Alexander teaches us that it is advantageous to increase the turns, and hence voltage, in the output coil. Section VI, (c).

3) Power increases with an increase in magnetic field strength. (A function of the wattage of the D.C. input E.M.F.)

4) When compared with conventional generators/alternators, there is no counter-torque on the rotor.

5) Very high efficiency; when compared with conventional generators. The recorded efficiency is: 125%.

For more information see Paul Brown's article titled — *The Moray Device and Hubbard Coil were Nuclear Batteries* on page 121.

Paul Brown's Project Work

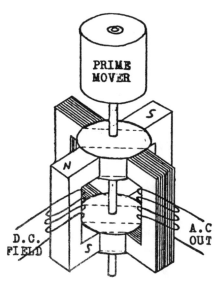

Note solid core in D.C. field coils.

DIMENSIONS AS ASSEMBLED

Variable Reluctance Generator

2

A Permanent Magnet Motor, 1269 A.D.

PERMANENT MAGNET MOTORS

Peregrinus (1269 A.D.) to Lee Bowman in 1954.

Peter Peregrinus is credited with the developement of the first known and recorded permanent magnet motor in 1269. His original work has been translated from Latin and the work is on file at the New York City Public Library.

The Peregrinus P.M.M. work remained dormant over the centuries until it was revived by Mr. Lee Bowman of California in 1954, who evolved a small scale working model.

The device consisted of three parallel shafts supported in bearings within end plates secured to a solid base plate. Three gears were secured at one end of each of the three shafts, at a two-to-one ratio, with one larger gear on the central shaft, as shown.

At the opposite end, three discs were secured to the shaft ends with one larger disc on the central shaft, and the two equal size smaller discs on the two, outer shafts. The discs were also fixed at a two-to-one ratio, the same as the gear ratios at the opposite shaft ends.

Eight Alnico rod permanent magnets were equally spaced on the one large disc, and four magnets each on the two smaller discs, so that they would coincide in position when the three discs were revolved. The elongated Alnico permanent magnets were placed on each of the discs so that they revolved parallel to the shafts, and their ends passed each other with a close air gap of about .005''.

When the discs were moved by hand, the magnets passing each other were so phased as to be synchronized at each passing position, as shown in the sketches.

The operation of the magnetic device required the positioning of a single cylindrical permanent magnet which was placed at an angle relative to the lower quadrant of the end discs, as shown. This single magnet acted as the actuator magnet which caused the rotation of the disc by unbalancing the magnetic forces of the three magnetic discs.

The Bowman magnetic motor was witnessed by several people including an electrical engineer who was impressed with its operation at the time of the demonstration. Although the Bowman device ahd received some exposure it never received any development interest and was eventually dismantled and destroyed, with no records made of its development potential.

The Bowman Permanent Magnet Motor

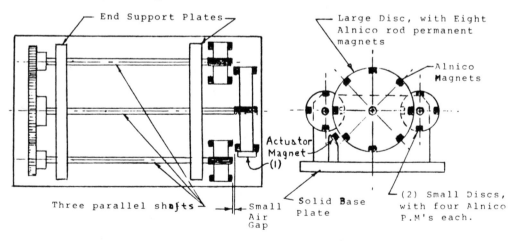

THE LETTER OF PETRUS PEREGRINUS

ON THE MAGNET, A.D. 1269

TRANSLATED BY

BROTHER ARNOLD, M.Sc.

PRINCIPAL OF LA SALLE INSTITUTE, TROY

WITH

INTRODUCTORY NOTICE

BY

BROTHER POTAMIAN, D.Sc.

PROFESSOR OF PHYSICS IN MANHATTAN
COLLEGE, NEW YORK

NEW YORK
McGRAW PUBLISHING COMPANY
MCMIV

INTRODUCTORY

THE magnetic lore of classic antiquity was scanty indeed, being limited to the attraction which the lodestone manifests for iron. Lucretius (99-55 B. C.), however, in his poetical dissertation on the magnet, contained in *De Rerum Natura*, Book VI.[1] recognizes magnetic repulsion, magnetic induction, and to some extent the magnetic field with its lines of force, for in verse 1040 he writes:

> Oft from the magnet, too, the steel recedes,
> Repelled by turns and re-attracted close.

And in verse 1085 :

> Its viewless, potent virtues men surprise ;
> Its strange effects, they view with wond'ring eyes

---

[1] With very few exceptions all the works referred to in this notice will be found in the Wheeler Collection in the Library of the American Institute of Electrical Engineers, New York.

THE LETTER OF PEREGRINUS

When without aid of hinges, links or springs
A pendant chain we hold of steely rings
Dropt from the stone—the stone the binding source—
Ring cleaves to ring and owns magnetic force :
Those held above, the ones below maintain,
Circle 'neath circle downward draws in vain
Whilst free in air disports the oscillating chain.

The poet Claudian (365-408 A. D.) wrote a short idyll on the attractive virtue of the lodestone and its symbolism ; St. Augustine (354-430), in his work *De Civitate Dei*, records the fact that a lodestone, held under a silver plate, draws after it a scrap of iron lying on the plate. Abbot Neckam, the Augustinian (1157-1217), distinguishes between the properties of the two ends of the lodestone, and gives in his *De Utensilibus*, what is perhaps the earliest reference to the mariner's compass that we have. Albertus Magnus, the Dominican (1193-1280), in his treatise, *De Mineralibus*, enumerates different kinds of natural magnets and states some of the properties commonly attributed to them ; the minstrel, Guyot de Provins, in a famous satirical poem, written about 1208, refers to the directive qual-

INTRODUCTORY

ity of the lodestone and its use in navigation, as
do also Cardinal de Vitry in his *Historia Orientalis* (1215-1220); Brunetto Latini, poet, orator
and philosopher, in his *Trésor des Sciences*, a veri-
table library, written in Paris in 1260; Raymond Lully, the Enlightened Doctor, in his
treatise, *De Contemplatione*, begun in 1272, and
Guido Guinicelli, the poet-priest of Bologna,
who died in 1276.

The authors of these learned works were too
busy with the pen to find time to devote to the
close and prolonged study of natural phenomena
necessary for fruitful discovery, and so had to content themselves with recording and discussing in
their tomes the scientific knowledge of their age
without making any notable additions to it.

But this was not the case with such contem-
poraries of theirs as Roger Bacon, the Franciscan, and his Gallic friend, Pierre de Maricourt,
commonly called Petrus Peregrinus, the subject
of the present notice, a man of academic culture
and of a practical rather than speculative turn of
mind. Of the early years of Peregrinus nothing

THE LETTER OF PEREGRINUS

is known save that he studied probably at the University of Paris, and that he graduated with the highest scholastic honors.   He owes his surname to the village of Maricourt, in Picardy, and the appellation Peregrinus, or Pilgrim, to his having visited the Holy Land as a member of one of the crusading expeditions of the time.

In 1269 we find him in the engineering corps of the French army then besieging Lucera, in Southern Italy, which had revolted from the authority of its French master, Charles of Anjou. To Peregrinus was assigned the work of fortifying the camp and laying mines as well as of constructing engines for projecting stones and fireballs into the beleaguered city.

It was in the midst of such warlike preoccupations that the idea seems to have occurred to him of devising a piece of mechanism to keep the astronomical sphere of Archimedes in uniform rotation for a definite time.   In the course of his work over the new motor, Peregrinus was gradually led to consider the more fascinating problem of perpetual motion itself with the result

INTRODUCTORY

that he showed, at least diagrammatically, and to his own evident satisfaction, how a wheel might be driven round forever by the power of magnetic attraction.

Elated over his imaginary success, Peregrinus hastened to inform a friend of his at home; and that his friend might the more readily comprehend the mechanism of the motor and the functions of its parts, he proceeds to set forth in a methodical manner all the properties of the lodestone, most of which he himself had discovered. It is a fortunate circumstance that this Picard friend of his was not a man learned in the sciences, otherwise we would probably never have had the remarkable exposition which Peregrinus gives of the phenomena and laws of magnetism. This letter of 3,500 words is the first great landmark in the domain of magnetic philosophy, the next being Gilbert's *De Magnete*, in 1600.

The letter was addressed from the trenches at Lucera, Southern Italy, in August, 1269, to Sigerus de Foucaucourt, his "amicorum intimus," the dearest of friends. A more enlightened friend,

THE LETTER OF PEREGRINUS

however, than the knight of Foucaucourt was
Roger Bacon, who held Peregrinus in the very
highest esteem, as the following glowing testi-
mony shows : " There are but two perfect math-
ematicians," wrote the English monk, " John of
London and Petrus de Maharne-Curia, a Picard."
Further on in his *Opus Tertium*, Bacon thus ap-
praises the merits of the Picard : " I know of
only one person who deserves praise for his work
in experimental philosophy, for he does not care
for the discourses of men and their wordy warfare, but quietly and diligently pursues the works
of wisdom.   Therefore, what others grope after
blindly, as bats in the evening twilight, this man
contemplates in all their brilliancy because he is
a master of experiment.   Hence, he knows all
natural science whether pertaining to medicine
and alchemy, or to matters celestial and terrestrial.   He has worked diligently in the smelting
of ores as also in the working of minerals ; he is
thoroughly acquainted with all sorts of arms and
implements used in military service and in hunting, besides which he is skilled in agriculture and

INTRODUCTORY

in the measurement of lands. It is impossible to write a useful or correct treatise in experimental philosophy without mentioning this man's name. Moreover, he pursues knowledge for its own sake; for if he wished to obtain royal favor, he could easily find sovereigns who would honor and enrich him."

This last statement is worthy of the best utterances of the twentieth century. Say what they will, the most ardent pleaders of our day for original work and laboratory methods cannot surpass the Franciscan monk of the thirteenth century in his denunciation of mere book learning or in his advocacy of experiment and research, while in Peregrinus, the mediævalist, they have Bacon's impersonation of what a student of science ought to be. Peregrinus was a hard worker, nor a mere theorizer, preferring, Procrustean-like, to make theory fit the facts rather than facts the theory; he was a brilliant discoverer who knew at the same time how to use his discoveries for the benefit of mankind; he was a pioneer of science and a leader in the progress of the world.

THE LETTER OF PEREGRINUS

An analysis of the " Epistola " shows that

(*a*) Peregrinus was the first to assign a definite position to the poles of a lodestone, and to give directions for determining which is north and which south ;

(*b*) He proved that unlike poles attract each other, and that similar ones repel ;

(*c*) He established by experiment that every fragment of a lodestone, however small, is a complete magnet, thus anticipating one of our fundamental laboratory illustrations of the molecular theory ;

(*d*) He recognized that a pole of a magnet may neutralize a weaker one of the same name, and even reverse its polarity ;

(*e*) He was the first to pivot a magnetized needle and surround it with a graduated circle, Figs. 2 and 3.'

(*f*) He determined the position of an object by its magnetic bearing as done to-day in compass surveying ; and

---

' It is probable that Flavio Gioja, an Italian pilot, some fifty years later, added the compass-card and attached it to the magnet.

INTRODUCTORY

(g) He introduced into his perpetual motion machine, Fig. 4, the idea of a magnetic motor, a clever idea, indeed, for a thirteenth century engineer.

This rapid summary will serve to show that the letter of Peregrinus is one of great interest in physics as well as in navigation and geodesy. For nearly three centuries, it lay unnoticed among the libraries of Europe, but it did not escape Gilbert, who makes frequent mention of it in his *De Magnete*, 1600; nor the illustrious Jesuit writers, Cabæus, who refers to it in his *Philosophia Magnetica*, 1629, and Kircher, who quotes from it in his *De Arte Magnetica*, 1641; it was well known to Jean Taisnier, the Belgian plagiarist, who transferred a great part of it verbatim to the pages of his *De Natura Magnetis*, 1562, without a word of acknowledgment. By this piece of fraud, Taisnier acquired considerable celebrity, a fact that goes to show the meritorious character of the work which he unscrupulously copied.

This memorable letter is divided into two

THE LETTER OF PEREGRINUS

parts : the first contains ten chapters on the general properties of the lodestone; the second has but three chapters, and shows how the author proposed to use a lodestone for the purpose of producing continuous rotation.

There are many manuscript copies of the letter in European libraries : the Bodleian has six ; the Vatican, two; Trinity College, Dublin, one; the Bibliothèque Nationale, Paris, one ; Leyden, Geneva and Turin, one each. The Leyden MS. has acquired special notoriety from a passage which appears near the end of it in which reference is made to magnetic declination and its value given : but Prof. W. Wenckebach, of The Hague, has shown' that the lines are spurious, having been interpolated in the manuscript in the early part of the sixteenth century.

The Leyden manuscript has also led some writers to believe in a fictitious author of the letter, one Peter Adsiger, or Petrus Adsigerus. As said above, Sigerus was the name of his countryman, to whom Peregrinus addressed his letter,

---

' Annali di Matematica Pura ed Applicata, 1865.

INTRODUCTORY

the *Epistola ad Sigerum*, from the trenches at Lucera, in August, 1269.

Magnetic declination was unknown to Peregrinus, else he would not have written the following words: " Wherever a man may be, he finds
the lodestone pointing to the heavens in accordance with the position of the meridian " (Chapter
X). Of course, the geographical meridian is the
one here meant, as the necessity of a distinct
magnetic meridian had not yet occurred to any
one.

Nor was this important magnetic element
known to Columbus when he sailed from the
shores of the Old World in 1492 as appears from
the surprise with which he noticed the deviation
of the needle from North as well as from the
consternation of his pilots. Columbus has the
unquestionable merit of being the first to observe
and record the change of declination with change
of place.

The first printed edition of the Epistola, now
very rare, was prepared by Achilles Gasser, a phy-
sician of Lindau, a man well versed in mathe-

THE LETTER OF PEREGRINUS

matics, astronomy, history and philosophy. The
work was printed in Augsburg in 1558. A copy
of this early print is among the treasures of the
Wheeler collection in the library of the American Institute of Electrical Engineers, New York.
It was from this text that the translation which
follows was made.

Besides the Latin edition of Gasser, 1558,
there is also that of Libri in his *Histoire des Sciences Mathématiques*, 1838 ; of Bertelli, 1868, and
Hellmann, 1898. Bertelli's is a learned and exhaustive work in which the Barnabite monk, sometimes called by mistake, Barnabita, instead of Bertelli, collates and compares the readings of the
two Vatican codices with other texts, adding copi-
ous references and explanatory notes. It appeared
in the *Bulletino di Bibliografia e di Storia delle Scienze
Matematiche e Fisiche* for 1868.

Of translations, we have that which Richard
Eden made from Taisnier's pirated extracts, the
first dated edition appearing in 1579. Cavallo's
Treatise on Magnetism, 1800, also contains some
of the more remarkable passages. The only com-

INTRODUCTORY

plete English translation that we have, appeared
in 1902 from the scholarly pen of Prof. Silvanus
P. Thompson, of London.   It is an *édition de luxe*
beautifully rubricated, but limited to 250 copies.
The translation was based on the texts of Gasser
and Hellmann, amended by reference to a manuscript in the author's possession, dated 1391.
We are informed that Mr. Fleury P. Mottelay,
of New York, the learned translator of Gilbert's
De Magnete, possesses a manuscript version by
Prof. Peirce, of Harvard, of the Paris codex, of
which he made a careful study in an endeavor to
decipher the illegible parts.

THE LETTER OF PEREGRINUS

PART I

—

CHAPTER I

PURPOSE OF THIS WORK

DEAREST OF FRIENDS:

AT your earnest request, I will now make known to you, in an unpolished narrative, the undoubted though hidden virtue of the lodestone, concerning which philosophers up to the present time give us no information, because it is characteristic of good things to be hidden in darkness until they are brought to light by application to public utility.   Out of affection for you, I will write in a simple style about things entirely unknown to the ordinary individual. Nevertheless I will speak only of the manifest properties of the lodestone, because this tract will form part of a work on the construction of philosophical instruments.   The disclosing of the

THE LETTER OF PEREGRINUS

hidden properties of this stone is like the art of
the sculptor by which he brings figures and seals
into existence.   Although I may call the matters
about which you inquire evident and of inesti-
mable value, they are considered by common
folk to be illusions and mere creations of the imagination.   But the things that are hidden from
the multitude will become clear to astrologers
and students of nature, and will constitute their
delight, as they will also be of great help to those
that are old and more learned.

CHAPTER II

QUALIFICATIONS OF THE EXPERIMENTER

YOU must know, my dear friend, that who-
ever wishes to experiment, should be acquainted with the nature of things, and should
not be ignorant of the motion of the celestial
bodies.   He must also be skilful in manipulation
in order that, by means of this stone, he may produce these marvelous effects.   Through his own
industry he can, to some extent, indeed, correct

THE LETTER OF PEREGRINUS

the errors that a mathematician would inevitably make if he were lacking in dexterity. Besides, in such occult experimentation, great skill is required, for very frequently without it the desired result cannot be obtained, because there are many things in the domain of reason which demand this manual dexterity.

CHAPTER III

CHARACTERISTICS OF A GOOD LODESTONE

THE lodestone selected must be distinguished by four marks—its color, homogeneity, weight and strength. Its color should be iron-like, pale, slightly bluish or indigo, just as polished iron becomes when exposed to the corroding atmosphere. I have never yet seen a stone of such description which did not produce wonderful effects. Such stones are found most frequently in northern countries, as is attested by sailors who frequent places on the northern seas, notably in Normandy, Flanders and Picardy. This stone should also be of homogeneous ma-

THE LETTER OF PEREGRINUS

terial ; one having reddish spots and small holes in it should not be chosen; yet a lodestone is hardly ever found entirely free from such blemishes.   On account of uniformity in its composition and the compactness of its innermost parts, such a stone is heavy and therefore more valuable.   Its strength is known by its vigorous attraction for a large mass of iron ; further on I will explain the nature of this attraction. If you chance to see a stone with all these characteristics, secure it if you can.

CHAPTER IV

HOW TO DISTINGUISH THE POLES OF A
LODESTONE

I WISH to inform you that this stone bears in itself the likeness of the heavens, as I will now clearly demonstrate. There are in the heavens two points more important than all others, because on them, as on pivots, the celestial sphere revolves : these points are called, one the arctic or north pole, the other the antarctic or south pole.   Similarly you must fully realize that in

THE LETTER OF PEREGRINUS

this stone there are two points styled respectively the north pole and the south pole. If you
are very careful, you can discover these two
points in a general way. One method for doing
so is the following: With an instrument with
which crystals and other stones are rounded let
a lodestone be made into a globe and then polished.   A needle or an elongated piece of iron
is then placed on top of the lodestone and a line
is drawn in the direction of the needle or iron,
thus dividing the stone into two equal parts.
The needle is next placed on another part of the
stone and a second median line drawn.   If desired, this operation may be performed on many
different parts, and undoubtedly all these lines
will meet in two points just as all meridian or
azimuth circles meet in the two opposite poles
of the globe.   One of these is the north pole,
the other the south pole.   Proof of this will be
found in a subsequent chapter of this tract.

A second method for determining these important points is this: Note the place on the
above-mentioned spherical lodestone where the
point of the needle clings most frequently and

THE LETTER OF PEREGRINUS

most strongly; for this will be one of the poles as discovered by the previous method. In order to determine this point exactly, break off a small piece of the needle or iron so as to obtain a fragment about the length of two fingernails; then put it on the spot which was found to be the pole by the former operation. If the fragment stands perpendicular to the stone, then that is, unquestionably, the pole sought; if not, then move the iron fragment about until it becomes so; mark this point carefully; on the opposite end another point may be found in a similar manner. If all this has been done rightly, and if the stone is homogeneous throughout and a choice specimen, these two points will be diametrically opposite, like the poles of a sphere.

CHAPTER V

HOW TO DISCOVER THE POLES OF A LODESTONE AND HOW TO TELL WHICH IS NORTH AND WHICH SOUTH

THE poles of a lodestone having been located in a general way, you will determine which is north and which south in the following man-

THE LETTER OF PEREGRINUS

ner :   Take a wooden vessel rounded like a plat-
ter or dish, and in it place the stone in such a
way that the two poles will be equidistant from
the edge of the vessel ; then place the dish in
another and larger vessel full of water, so that
the stone in the first-mentioned dish may be like
a sailor in a boat.   The second vessel should be
of considerable size so that the first may resemble
a ship floating in a river or on the sea.   I insist
upon the larger size of the second vessel in order
that the natural tendency of the lodestone may
not be impeded by contact of one vessel against
the sides of the other.   When the stone has been
thus placed, it will turn the dish round until the
north pole lies in the direction of the north pole
of the heavens, and the south pole of the stone
points to the south pole of the heavens.   Even
if the stone be moved a thousand times away from
its position, it will return thereto a thousand
times, as by natural instinct.   Since the north
and south parts of the heavens are known, these
same points will then be easily recognized in
the stone because each part of the lodestone will
turn to the corresponding one of the heavens.

THE LETTER OF PEREGRINUS

CHAPTER VI

HOW ONE LODESTONE ATTRACTS ANOTHER

WHEN you have discovered the north and the south pole in your lodestone, mark them both carefully, so that by means of these indentations they may be distinguished whenever necessary. Should you wish to see how one lodestone attracts another, then, with two lodestones selected and prepared as mentioned in the preceding chapter, proceed as follows: Place one in its dish that it may float about as a sailor in a skiff, and let its poles which have already been determined be equidistant from the horizon, i. e., from the edge of the vessel. Taking the other stone in your hand, approach its north pole to the south pole of the lodestone floating in the vessel; the latter will follow the stone in your hand as if longing to cling to it. If, conversely, you bring the south end of the lodestone in your hand toward the north end of the floating lodestone, the same phenomenon will occur; namely, the floating lodestone will follow the one in your hand. Know then that this is the law: the north

THE LETTER OF PEREGRINUS

pole of one lodestone attracts the south pole of another, while the south pole attracts the north. Should you proceed otherwise and bring the north pole of one near the north pole of another, the one you hold in your hand will seem to put the floating one to flight.   If the south pole of one is brought near the south pole of another, the same will happen.   This is because the north pole of one seeks the south pole of the other, and therefore repels the north pole.   A proof of this is that finally the north pole becomes united with the south pole.   Likewise if the south pole is stretched out towards the south pole of the floating lodestone, you will observe the latter to be repelled, which does not occur, as said before, when the north pole is extended towards the south.   Hence the silliness of certain persons is manifest, who claim that just as scammony attracts jaundice on account of a similarity between them, so one lodestone attracts another even more strongly than it does iron, a fact which they suppose to be false although really true as shown by experiment.

THE LETTER OF PEREGRINUS

CHAPTER VII

HOW IRON TOUCHED BY A LODESTONE TURNS
TOWARDS THE POLES OF THE WORLD

IT is well known to all who have made the experiment, that when an elongated piece of iron has touched a lodestone and is then fastened to a light block of wood or to a straw and made float on water, one end will turn to the star which has been called the Sailor's star because it is near the pole; the truth is, however, that it does not point to the star but to the pole itself. A proof of this will be furnished in a following chapter. The other end of the iron will point in an opposite direction. But as to which end of the iron will turn towards the north and which to the south, you will observe that that part of the iron which has touched the south pole of the lodestone will point to the north and conversely, that part which had been in contact with the north pole will turn to the south. Though this appears marvelous to the uninitiated, yet it is known with certainty to those who have tried the experiment.

THE LETTER OF PEREGRINUS

CHAPTER VIII

HOW A LODESTONE ATTRACTS IRON

I F you wish the stone, according to its natural
desire, to attract iron, proceed as follows:
Mark the north end of the iron and towards
this end approach the south pole of the stone,
when it will be found to follow the latter. Or,
on the contrary, to the south part of the iron
present the north pole of the stone and the latter will attract it without any difficulty. Should
you, however, do the opposite, namely, if you
bring the north end of the stone towards the
north pole of the iron, you will notice the iron
turn round until its south pole unites with the
north end of the lodestone. The same thing
will occur when the south end of the lodestone
is brought near the south pole of the iron.
Should force be exerted at either pole, so that
when the south pole of the iron is made touch
the south end of the stone, then the virtue in
the iron will be easily altered in such a manner
that what was before the south end will now
become the north and conversely. The cause is

that the last impression acts, confounds, or count-
eracts and alters the force of the original movement.

CHAPTER IX

WHY THE NORTH POLE OF ONE LODESTONE ATTRACTS THE SOUTH POLE OF ANOTHER AND VICE VERSA

AS already stated, the north pole of one lodestone attracts the south pole of another
and conversely; in this case the virtue of the
stronger becomes active, whilst that of the weaker
becomes obedient or passive. I consider the following to be the cause of this phenomenon: the
active agent requires a passive subject, not merely
to be joined to it, but also to be united with it,
so that the two make but one by nature. In the
case of this wonderful lodestone this may be
shown in the following manner: Take a lodestone which you may call $A D$, in which A is
the north pole and D the south; cut this stone
into two parts, so that you may have two distinct

THE LETTER OF PEREGRINUS

stones; place the stone having the pole A so that it may float on water and you will observe that A turns towards the north as before; the breaking did not destroy the properties of the parts of the stone, since it is homogeneous; hence it follows that the part of the stone at the point of fracture, which may be marked B, must be a south pole; this broken part of which we are now speaking may be called $A B$. The other, which contains D, should then be placed so as to float on water, when you will see D point towards the south because it is a south pole; but the other end at the point of fracture, lettered C, will be a north pole; this stone may now be named $C D$. If we consider the first stone as the active agent, then the second, or $C D$, will be the passive subject. You will also notice that the ends of the two stones which before their separation were together, after breaking will become one a north pole and the other a south pole. If now these same broken portions are brought near each other, one will attract the other, so that they will again be

THE LETTER OF PEREGRINUS

joined at the points B and C, where the fracture occurred. Thus, by natural instinct, one single stone will be formed as before. This may be demonstrated fully by cementing the parts together, when the same effects will be produced as before the stone was broken. As you will perceive from this experiment, the active agent desires to become one with the passive subject because of the similarity that exists between them. Hence C, being a north pole, must be brought close to B, so that the agent and its subject may form one and the same straight line in the order A B, C D and B and C being at the same point. In this union the identity of the extreme parts is retained and preserved just as they were at first; for A is the north pole in the entire line as it was in the divided one; so also D is the south pole as it was in the divided passive subject, but B and C have been made effectually into one. In the same way it happens that if A be joined to D so as to make the two lines one, in virtue of this union due to attraction in the order C D A B, then A and D

THE LETTER OF PEREGRINUS

will constitute but one point, the identity of the
extreme parts will remain unchanged just as they
were before being brought together, for C is a
north pole and B a south, as during their separation.   If you proceed in a different fashion,
this identity or similarity of parts will not be
preserved; for you will perceive that if C, a
north pole, be joined to A, a north pole, contrary to the demonstrated truth, and from these
two lines a single one, $B\ A\ C\ D$, is formed, as
D was a south pole before the parts were united,
it is then necessary that the other extremity
should be a north pole, and as B is a south pole,
the identity of the parts of the former similarity
is destroyed.   If you make B the south pole as
it was before they united, then D must become
north, though it was south in the original stone;
in this way neither the identity nor similarity
of parts is preserved.   It is becoming that when
the two are united into one, they should bear
the same likeness as the agent, otherwise nature
would be called upon to do what is impossible.
The same incongruity would occur if you were

THE LETTER OF PEREGRINUS

to join *B* with *D* so as to make the line *A B D C*, as is plain to any person who reflects a moment. Nature, therefore, aims at being and also at acting in the best manner possible; it selects the former motion and order rather than the second because the identity is better preserved. From all this it is evident why the north pole attracts the south and conversely, and also why the south pole does not attract the south pole and the north pole does not attract the north.

CHAPTER X

AN INQUIRY INTO THE CAUSE OF THE NATURAL
VIRTUE OF THE LODESTONE

CERTAIN persons who were but poor investigators of nature held the opinion that the force with which a lodestone draws iron, is found in the mineral veins themselves from which the stone is obtained; whence they claim that the iron turns towards the poles of the earth, only because of the numerous iron mines found there. But such persons are ignorant of the fact that in

THE LETTER OF PEREGRINUS

many different parts of the globe the lodestone
is found; from which it would follow that the iron
needle should turn in different directions according to the locality ; but this is contrary to experience. Secondly, these individuals do not seem to
know that the places under the poles are unin-
habitable because there one-half the year is day
and the other half night.   Hence it is most silly
to imagine that the lodestone should come to us
from such places. Since the lodestone points to the
south as well as to the north, it is evident from
the foregoing chapters that we must conclude
that not only from the north pole but also from
the south pole rather than from the veins of the
mines virtue flows into the poles of the lodestone.
This follows from the consideration that wherever a man may be, he finds the stone pointing
to the heavens in accordance with the position
of the meridian; but all meridians meet in the
poles of the world ; hence it is manifest that
from the poles of the world, the poles of the
lodestone receive their virtue.   Another necessary consequence of this is that the needle does

THE LETTER OF PEREGRINUS

not point to the pole star, since the meridians do not intersect in that star but in the poles of the world. In every region, the pole star is always found outside the meridian except twice in each complete revolution of the heavens. From all these considerations, it is clear that the poles of the lodestone derive their virtue from the poles of the heavens. As regards the other parts of the stone, the right conclusion is, that they obtain their virtue from the other parts of the heavens, so that we may infer that not only the poles of the stone receive their virtue and influence from the poles of the world, but likewise also the other parts, or the entire stone from the entire heavens. You may test this in the following manner : A round lodestone on which the poles are marked is placed on two sharp styles as pivots having one pivot under each pole so that the lodestone may easily revolve on these pivots. Having done this, make sure that it is equally balanced and that it turns smoothly on the pivots. Repeat this several times at different hours of the day and always with the utmost

THE LETTER OF PEREGRINUS

care.   Then place the stone with its axis in the meridian, the poles resting on the pivots. Let it be moved after the manner of bracelets so that the elevation and depression of the poles may equal the elevation and depressions of the poles of the heavens of the place in which you are experimenting.   If now the stone be moved according to the motion of the heavens, you will be delighted in having discovered such a wonderful secret ; but if not, ascribe the failure to your own lack of skill rather than to a defect in nature.   Moreover, in this position I consider the strength of the lodestone to be best preserved. When it is placed differently, i. e., not in the meridian, I think its virtue is weakened or obscured rather than maintained. With such an instrument you will need no timepiece, for by it you can know the ascendant at any hour you please, as well as all other dispositions of the heavens which are sought for by astrologers.

THE LETTER OF PEREGRINUS

PART II

CHAPTER I

THE CONSTRUCTION OF AN INSTRUMENT FOR MEAS-
URING THE AZIMUTH OF THE SUN
THE MOON OR ANY STAR
ON THE HORIZON

HAVING fully examined all the properties of the lodestone and the phenomena connected therewith, let us now come to those instruments which depend for their operation on the knowledge of those facts.   Take a rounded lodestone,[1] and after determining its poles in the manner already mentioned, file its two sides so that it becomes elongated at its poles and occupies less space.   The lodestone prepared in this wise is then enclosed within two capsules after the fashion of a mirror.   Let these capsules be so joined together that they cannot be sepa-

---
[1] A terrella, or earthkin.

THE LETTER OF PEREGRINUS

rated and that water cannot enter ; they should be made of light wood and fastened with cement suited to the purpose.   Having done this, place them in a large vessel of water on the edges of which the two parts of the world, i. e., the north and south points, have been found and marked.   These points may be united by a thread stretched across from north to south. Then float the capsules and place a smooth strip of wood over them in the manner of a diameter.   Move the strip until it is equally distant from the meridian-line, previously determined and marked by a thread, or else until it coincides therewith.   Then mark a line on the capsules according to the position of the strip, and this will indicate forever the meridian of that place.   Let this line be divided at its middle by another cutting it at right angles, which will give the east and west line ; thus the four cardinal points will be determined and indicated on the edge of the capsules.   Each quarter is to be subdivided into 90 parts, making 360 in the circumference of the capsules. Engrave these divi-

THE LETTER OF PEREGRINUS

sions on them as usually done on the back of
an astrolabe.   On the top or edge of the capsules thus marked place a thin ruler like the
pointer on the back of the astrolabe ; instead of
the sights attach two perpendicular pins, one at
each end.   If, therefore, you desire to take the
azimuth of the sun, place the capsules in water
and let them move freely until they come to
rest in their natural position.   Hold them firmly
in one hand, while with the other you move the
ruler until the shadow of the pins falls along the
length of the ruler; then the end of the ruler
which is towards the sun will indicate the azimuth of the sun.   Should it be windy, let the
capsules be covered with a suitable vessel until
they have taken their position north and south.
The same method, namely, by sighting, may be
followed at night for determining the azimuth
of the moon and stars ; move the ruler until the
ends of the pins are in the same line with the
moon or star; the end of the ruler will then indicate the azimuth just as in the case of the sun.
By means of the azimuth may then be deter-

THE LETTER OF PEREGRINUS

mined the hour of the day, the ascendant, and all those other things usually determined by the astrolabe.   A form of the instrument is shown in the following figure.

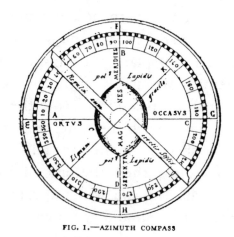

FIG. I.—AZIMUTH COMPASS

CHAPTER II

THE CONSTRUCTION OF A BETTER INSTRUMENT
FOR THE SAME PURPOSE

IN this chapter I will describe the construction of a better and more efficient instrument. Select a vessel of wood, brass or any solid material you like, circular in shape, moderate in

THE LETTER OF PEREGRINUS

size, shallow but of sufficient width, with a cover
of some transparent substance, such as glass or
crystal; it would be even better to have both
the vessel and the cover transparent. At the
centre of this vessel fasten a thin axis of brass
or silver, having its extremities in the cover
above and the vessel below. At the middle of
this axis let there be two apertures at right angles to each other; through one of them pass
an iron stylus or needle, through the other a silver or brass needle crossing the iron one at right
angles. Divide the cover first into four parts
and subdivide these into 90 parts, as was mentioned in describing the former instrument.
Mark the parts north, south, east and west. Add
thereto a ruler of transparent material with pins
at each end. After this bring either the north
or the south pole of a lodestone near the cover
so that the needle may be attracted and receive
its virtue from the lodestone. Then turn the
vessel until the needle stands in the north and
south line already marked on the instrument;
after which turn the ruler towards the sun if

THE LETTER OF PEREGRINUS

day-time, and towards the moon and stars at night, as described in the preceding chapter. By means of this instrument you can direct your course towards cities and islands and any other

FIG. 2.—DOUBLE-PIVOTED NEEDLE

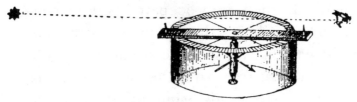

FIG. 3.—PIVOTED COMPASS

place wherever you may wish to go by land or sea, provided the latitude and longitude of the places are known to you. How iron remains suspended in air by virtue of the lodestone, I will explain in my book on the action of mir-

THE LETTER OF PEREGRINUS

rors. Such, then, is the description of the instrument illustrated below. (See Figs. 2 and 3.)

CHAPTER III

THE ART OF MAKING A WHEEL OF
PERPETUAL MOTION

IN this chapter I will make known to you the
construction of a wheel which in a remarkable manner moves continuously. I have seen
many persons vainly busy themselves and even
becoming exhausted with much labor in their
endeavors to invent such a wheel. But these in-
variably failed to notice that by means of the virtue or power of the lodestone all difficulty can be
overcome. For the construction of such a wheel,
take a silver capsule like that of a concave mirror, and worked on the outside with fine carv-
ing and perforations, not only for the sake of
beauty, but also for the purpose of diminishing
its weight. You should manage also that the
eye of the unskilled may not perceive what is
cunningly placed inside. Within let there be

THE LETTER OF PEREGRINUS

iron nails or teeth of equal weight fastened to
the periphery of the wheel in a slanting direction, close to one another so that their distance
apart may not be more than the thickness of a
bean or a pea ; the wheel itself must be of uniform weight throughout.   Fasten the middle of
the axis about which the wheel revolves so that
the said axis may always remain immovable. Add
thereto a silver bar, and at its extremity affix a
lodestone placed between two capsules and prepared in the following way : When it has been
rounded and its poles marked as said before, let
it be shaped like an egg ; leaving the poles un-
touched, file down the intervening parts so that
thus flattened and occupying less space, it may
not touch the sides of the capsules when the
wheel revolves. Thus prepared, let it be attached
to the silver rod just as a precious stone is placed
in a ring ; let the north pole be then turned towards the teeth or cogs of the wheel somewhat
slantingly so that the virtue of the stone may not
flow diametrically into the iron teeth, but at a
certain angle ; consequently when one of the

THE LETTER OF PEREGRINUS

teeth comes near the north pole and owing to
the impetus of the wheel passes it, it then approaches the south pole from which it is rather
driven away than attracted, as is evident from the
law given in a preceding chapter. Therefore such
a tooth would be constantly attracted and con-

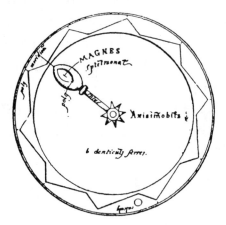

FIG. 4.—PERPETUAL MOTION WHEEL

stantly repelled. In order that the wheel may
do its work more speedily, place within the box a
small rounded weight made of brass or silver of
such a size that it may be caught between each
pair of teeth; consequently as the movement of

the wheel is continuous in one direction, so the fall of the weight will be continuous in the other. Being caught between the teeth of a wheel which is continuously revolving, it seeks the centre of the earth in virtue of its own weight, thereby aiding the motion of the teeth and preventing them from coming to rest in a direct line with the lode-stone. Let the places between the teeth be suit-ably hollowed out so that they may easily catch the body in its fall, as shown in the diagram above. (Fig. 4.)

Farewell: finished in camp at the siege of Lucera on the eighth day of August, Anno Dom-ini MCCLXIX.

EARLY REFERENCES TO THE MARINER'S COMPASS

THE following are the passages referred to in the introductory notice:

Abbot Neckam (1157-1217), in his *De Naturis Rerum*,
writes:

"The sailors, moreover, as they sail over the sea, when in
cloudy weather they can no longer profit by the light of the sun,
or when the world is wrapped up in the darkness of the shades
of night and they are ignorant to what point their ship's course
is directed, these mariners touch the lodestone with a needle,
which (the needle) is whirled round in a circle until when its
motion ceases, its point looks direct to the north. (*Cuspis
ipsius septentrionalem plagam respiciat.*)"

In his *De Utensilibus*, we read:

"Among other stores of a ship, there must be a needle
mounted on a dart (*habeat etiam acum jaculo superpositam*)
which will oscillate and turn until the point looks to the north,
and the sailors will thus know how to direct their course when

THE LETTER OF PEREGRINUS

the pole star is concealed through the troubled state of the atmosphere."[1]

Alexander Neckam was born at St. Albans in 1157, joined the Augustinian Order and taught in the University of Paris from 1180 to 1187, after which he returned to England to take charge of a College of his Order at Dunstable.   He was elected Abbot of Cirencester in 1213 and died at Kemsey, near Worcester, in 1217.

The satirical poem of Guyot de Provins, written about 1208, contains the following passage:

> The mariners employ an art which cannot deceive,
> By the property of the lodestone,
> An ugly stone and brown,
> To which iron joints itself willingly
> They have; they attend to where it points
> After they have applied a needle to it ;
> And they lay the latter on a straw
> And put it simply in the water
> Where the straw makes it float.
> Then the point turns direct
> To the star with such certainty
> That no man will ever doubt it,
> Nor will it ever go wrong.
> When the sea is dark and hazy,
> That one sees neither star nor moon,
> Then they put a light by the needle
> And have no fear of losing their way.
> The point turns towards the star ;

---

[1] The Chronicles and Memoirs of Great Britain and Ireland during the Middle Ages, by Thomas Wright (1863).

NOTES

And the mariners are taught
To follow the right way.
It is an art which cannot fail.

Provins, from which Guyot took his surname, was a small town in the vicinity of Paris.

Cardinal Jacques de Vitry, in his *Historia Orientalis*, Cap. 89, writes:

"An iron needle, after having been in contact with the lodestone, turns towards the north star, so that it is very necessary for those who navigate the seas."

Jacques de Vitry was born at Argenteuil, near Paris, joined the fourth crusade, became Bishop of Ptolemais, and died in Rome in 1244. He wrote his "Description of Palestine," which forms the first book of his *Historia Orientalis*, in the East, between 1215 and 1220.

Albertus Magnus (1193-1280) in his *De Mineralibus*, Lib. II., Tract 3, Cap. 6, writes:

"It is the end of the lodestone which makes the iron that touched it turn to the north (*ad zoron*) and which is of use to mariners; but the other end of the needle turns toward the south (*ad aphron*)."

This illustrious Bavarian schoolman joined the Dominican Order in his youth, lectured to great audiences in Cologne, became bishop of Ratisbonne in 1260, and died in 1280. Thomas Aquinas the greatest of schoolmen, was among his pupils.

THE LETTER OF PEREGRINUS

In the Spanish code of laws, begun in 1256, during the reign of Alfonso el Sabio, and known as *Las Siete Partidas*, we read:

"Just as mariners are guided during the night by the needle, which replaces for them the shores and pole star alike, by showing them the course to pursue both in fair weather and foul, so those who are called upon to advise the King must always be guided by a spirit of justice."

Brunetto Latini, in his *Trésor des Sciences*, 1260, writes:

"The sailors navigate the seas guided by the two stars called the tramontanes, and each of the two parts of the lodestone directs the end of the needle to the star to which that part itself turns."

Brunetto Latini (1230-1294) was a man of great eminence in the thirteenth century; Dante was among his pupils at Florence. For political reasons, he removed to Paris, where he wrote his *Trésor* and also his *Tesoretto*. He visited Roger Bacon at Oxford about 1260.

In his treatise *De Contemplatione*, begun in 1272, Raymond Lully writes:

"As the needle, after having touched the lodestone, turns to the north, so the mariner's needle (*acus nautica*) directs them over the sea."

Lully was born at Palma in the Island of Majorca in 1236; he joined the Third Order of St. Francis, dying in 1315.

NOTES

Ristoro d'Arezzo, in his *Libro della Composizione del Mundo*, written in 1282, has the following:

"Besides this, there is the needle which guides the mariner, and which is itself directed by the star called the tramontane." [1]

The following metrical translation of a poem by Guido Guinicelli, an Italian priest, 1276, is from the pen of Dr. Park Benjamin, of New York:

> In what strange regions 'neath the polar star
> May the great hills of massy lodestone rise,
> Virtue imparting to the ambient air
> To draw the stubborn iron ; while afar
> From that same stone, the hidden virtue flies
> To turn the quivering needle to the Bear
> In splendor blazing in the Northern skies.

The above extracts show that the directive property of the magnetic needle was well known in England, France, Germany, Spain and Italy in the thirteenth century.   In the passage from Neckam, the *acum jaculo superpositam* has been construed by some to mean a form of pivoted needle, while in the letter of Peregrinus, 1269, the double pivoted form is clearly described.

---

[1] The pole-star was thus named in the south of France and the north of Italy because seen beyond the mountains (the Alps).

The oldest known writings......
...on *Alternate Energy!*

The *quest for energy* has occupied the efforts of the human race, since Prometheus stole fire from the Gods and then delivered it to mankind!   Petrus Peregrinus, was an engineer in the engineering corps of the French army during their seige of Lucera in southern Italy (1269AD).

In the midst of these hostilities, Petrus turned his thoughts to magnetism and its peculiar properties. Fortunately, Petrus was a scholar as well as a gentleman, and placed his thoughts on parchment where they laid undisturbed for hundreds of years in the libraries of Europe.

After its discovery, this letter of 3500 words was deemed the *first great landmark* **in the domain of magnetic philosophy....** the next being Gilbert's writing on magnetism--*De Magnete*--in 1690. Gilbert, himself, often refers to Petrus's letter in his treatise.

Now, for the first time since the turn of the century, Petrus Peregrinus's letter is being made available. Whether you are a historian, scientist or engineer, you will find this ancient document a unique and interesting perspective on a somewhat controversial topic..... *perpetual motion!*

Don't Delay..... Get your Copy Today!

$6.95US

3
Non-Conventional Energy and Propulsion Methods

NON-CONVENTIONAL ENERGY
AND PROPULSION METHODS

Thomas Valone, M.A., P.E.

Proceedings of the Intersociety Energy Conversion Engineering Conference

1991

Integrity Research Institute
1377 K Street NW, Suite 204
Washington, DC 20005

NON-CONVENTIONAL ENERGY AND PROPULSION METHODS

Thomas Valone, P.E.

Integrity Research Institute

1377 K Street NW, Suite 204

Washington, DC 20005

ABSTRACT

From the disaster of the Space Shuttle, Challenger, to the Kuwaiti oil well fires, we are reminded constantly of our dependence on dangerous, combustible fuels for energy and propulsion. Over the past ten years, there has been a considerable production of new and exciting inventions which defy conventional analysis. The term "non-conventional" was coined in 1980 by a Canadian engineer to designate a separate technical discipline for this type of endeavor. Since then, several conferences have been devoted solely to these inventions. Integrity Research Corp., an affiliate of the Institute, has made an effort to investigate each viable product, develop business plans for several to facilitate development and marketing, and, in some cases, assign an engineering student intern to building a working prototype. Each inventor discussed in this presentation has produced a unique device for "free energy" generation or highly efficient force production. Included in this paper is also a short summary for non- specialists explaining the physics of free energy generation along with a working definition. The

Proc. Intersoc. Energy Conver. Eng. Conference, 1991

main topics of discussion include: space power, inertial propulsion, kinetobaric force, magnetic motors, thermal fluctuations, over-unity heat pumps, ambient temperature superconductivity and nuclear battery.

INTRODUCTION

Non-Conventional Energy

Over ten years ago, I was privileged to present a paper at Dr. Hans Nieper's first German Symposium of Gravitational Field Energy which resulted in an article describing it for *Energy Unlimited* magazine [1]. The conference was, I believe, without precedence. Dr. Nieper wrote about the conference in a book he called *Conversion of Gravity Field Energy* [2]. He also revised the book into a later edition he called *Dr. Nieper's Revolution in Technology, Medicine and Society* [3]. Dr. Nieper introduced so many revolutionary scientists at that one conference that several people on this side of the ocean invited them to attend a few similar conferences here in North America. Reference is made to the **First, Second and Third International Symposia On Non-Conventional Energy Technology** (ISONCET) held about a year apart in the U.S. and Canada by the **Planetary Association for Clean Energy** (PACE) [4]. George Hathaway, who was the Chairman of the First ISONCET, even coined the new word "non-conventional" to describe the innovative effort he saw occurring [5]. Since then, the phrase has retained its value and meaning, demonstrated by the fact that even the U.S. Army picked up the phrase to use in its own SBIR solicitations! Starting in the early 1980's, the United States Psychotronics Association [6] also became a rich and vital arena for free energy (definition to follow) presentations and debates by notable Ph.D's. In 1984, the International Tesla Society (ITS) picked up where the three symposia left off and has sponsored non- conventional energy technology (NCET) conferences every two years since then [7]. In 1986, AZ Industries held an NCET conference, entitled "Meeting of the Minds" in Temecula, CA to gather more speakers and prototypes together [8]. In 1989, the Swiss joined the movement, with the Swiss Association for Free Energy (SAFE), and sponsored their own conference which was very well attended and simultaneously translated in English and German for the audience [9]. Since this author was an invited speaker at the conferences mentioned above, the article serves as a firsthand summary and primer for those who may have missed this small Energy Revolution.

Whenever NCET generators and propulsion methods are suggested, visions of perpetual motion, levitation, antigravity, and UFO's are conjured up as well. The history of NCET reads like a detective novel, full of intrigue, persecution, suspense, high stakes, secrecy, sell-outs, geniuses, criminals, fakes and lunatics.

Starting with a book like *The Sea of Energy, in Which the Earth Floats*, by T. Henry Moray [10], one can get a feeling for the excitement surrounding these inventions, especially when their discoveries have been replicated, as Moray's has (see Nuclear Battery section). Without going into an extensive historical review, the reader is referred to Cadake Industries [4], High Energy Enterprises [7], and *Energy Unlimited* Publications [11] for more detailed information. The rest of this article will center on new viable methods for energy generation and propulsion.

Since the phrase "free energy" appears in much of the NCET literature, it is important to attempt a popular definition before proceeding any further. In thermodynamics, Gibbs free energy or Hemholtz free energy is really the capacity of a system to perform work, after temperature, entropy and enthalpy are considered. In the NCET tradition, free energy has the same meaning as long as the energy is *free of charge*! Since this may be hard to pin down, a more formal definition would be "any energy generation method that produces a net energy output which exceeds the total energy input by some measurable quantity, thus activating some potential energy in the environment." Researchers

Proc. Intersoc. Energy Conver. Eng. Conference, 1991

look for the "over-unity" ratio when calculating energy output versus input as a proof of free energy. Examples actually are already in use: ocean thermal energy conversion (OTEC), over-unity heat pumps and even solar and wind energy. However, many examples of interest to free energy enthusiasts are the NCET type. We can define NCET as "any unusual or unique method for energy generation which anticipates or necessitates a further development in theoretical physics." While most are like OTEC, the best NCET invention would be similar to a solar cell, where the over-unity ratio is infinite since <u>no energy investment is needed</u> to obtain the electrical energy output. To the engineer who knows how to measure watts per sq. cm. for sunlight, this may be a confusing calculation of energy conversion. The best answer is that the use of a solar cell would look just like magical free energy to a 19th century scientist. NCET over-unity devices look the same way to us until future physicists take for granted zero- point energy densities, neutrino flux densities, etc. along with their corresponding energy transducers. NCET inventions can be grouped with *alternative energy* as well, since ecological principles guide the developers to analyze any biological and environmental impact, great or small. NCET examples are to be found throughout the remainder of this paper.

"When nature is eventually seen as refusing to express itself in the accepted language, the crisis explodes with the kind of violence that results from a breach of confidence" says Prigogine and Stengers [12]. Thomas Kuhn, in *The Structure of Scientific Revolutions* describes a clear pattern that can be witnessed when paradigm shifts are about to occur. A vital part of the shift is when the old language and theory does not explain the new phenomena. Many of us believe that we are experiencing that today with NCET inventions. A good example that comes to mind is the spectacular Hutchison Effect captured on film by engineer Hathaway who also reproduced the experiments with two Van de Graffs and a Tesla coil, levitating a 20 lb. vise, along with lighter objects, reported to the Third ISONCET.

The Integrity Research Institute

Since a real energy revolution is still waiting to happen, *a decade after* the First ISONCET, the Integrity Research Institute was formed to scientifically research and develop viable NCET inventions with private support for public benefit. Disasters such as the Challenger Space Shuttle, compared to an overgrown roman candle, and the Mideast war, reclaiming oil reserves, have driven home the primitive state of energy and propulsion in which we find ourselves as every major city chokes in its self- created smog. NCET research is no longer a speculative curiosity. It is a necessity required to provide clean transportation and avert the next energy crisis when conventional energy becomes scarce and unaffordable.

The Institute's purpose centers on energy usage in its broadest sense, researching non-conventional energy and transportation (NEXT Division), energy medicine, and the biological effects of energy distribution, such as electromagnetic pollution. Integrity Research Institute is nonprofit and devoted to research and public education.

What follows is an overview of the many promising areas of the NEXT investigation, without attempting to be comprehensive in scope. The basic principles of non-conventional energy and propulsion, along with the concepts behind some amazing inventions are presented

POWER AND ENERGY

Space Power

Proc. Intersoc. Energy Conver. Eng. Conference, 1991

Now that Biosphere II is testing the ability of a completely enclosed human/plant ecosystem to support itself, with the landing on Mars in mind, we need to think seriously about real space power. The limitations of our present energy and propulsion technology become apparent when one examines the Space Shuttle. The Space Shuttle attached to its fully fueled booster weighs 4.5 million pounds at liftoff. However, its payload capacity is 65,000 pounds, which is only 1.5% of its weight. Even more astonishing than this is the fact that the 3 million pounds of weight is lost in the attempt to deliver that 65,000 pound payload [13]. In other words, about 70% of its total weight is left behind! Although the attempt is made to recover the booster cylinders, even with their leaky seals, we still must examine the cost to the taxpayer for such a wasteful propulsion system. Moreover, NASA experts agree, we will never get to Mars or to the rest of the solar system, with such a primitive technology.

Solar, chemical and thermal energy sources power a spacecraft during its mission. All of the energy conversions that are needed in a modern spacecraft create energy losses that must be dissipated. The chief danger of any electrical energy system in space, especially in low earth orbit, is the space plasma of ionized gases that causes arcing and voltage breakdown at a threshold of only **800 volts**. This is a serious limitation on the space station voltage as well as an interplanetary spacecraft. Skylab was designed around a 100 volt DC system while the Space Shuttle is entirely designed around a 24 volt DC environment [13]. Some NCET inventions are compatible with space power requirements.

Energy Conversion

The various methods of energy conversion are summarized in the Energy Input and Output Technologies chart from the textbook, *Space*

Power (a copy of the chart is available from the Institute). We see no conversion method for chemical to mechanical and vice versa. We also do not see gravitational, tachyon or magnetic field energy listed. Are there reasons for giving such non-conventional energy any consideration? A brief tutorial will help illustrate the possibilities:

In most generators and motors the armature magnets are electromagnets. An electromagnet requires electricity and therefore energy to produce its magnetic field. It normally has some electrical resistance which requires a dissipation of energy per time. This energy per time or **power** (i.e. "watts") is then maintained for a period of time which gives us the product of power and time or **work** (i.e. "kilowatt-hours"). This last definition of work, however, is equal to energy). In other words, we conclude that **Energy = Work.** Whether it is mechanical, thermal, fluidic, nuclear, electromagnetic, or gravitational energy, once the source and sink are connected through our transducer (i.e., generator), energy is harnessed for useful work.

Superconductors

Now, to produce a continuous, nondissipative, macroscopic flow of electrons, and therefore a perpetual magnetic field, we can simply cryogenically cool an electrical circuit, made of the appropriate metals, until it is superconductive. Today, the temperature for that transition is getting closer to room temperature each day with new breakthroughs in superconductive materials. Below the point of transition, the voltage supply can be *disconnected* and the perpetual motion of the electrons is established as the current continues to flow. Magnetic fields created in such a manner have been maintained *for several years* with no further input of energy. In Buffalo, New York, Roswell Park Memorial Institute has a few superconducting magnets that were energized in the late 1970's and have not required re- energizing since then. Varian As-

Proc. Intersoc. Energy Conver. Eng. Conference, 1991

sociates, who manufactures the superconducting magnets, expected to sell a lot of electrical energizers with their magnets but the magnets have been behaving much better than anticipated. The basic circuit diagram that achieves **perpetual motion**, thanks to superconductivity, is a single loop where is the battery starting the current in the loop is switched out of the circuit once energized.

Spinning Electrons Create Ferromagnetism

Why does this superconducting circuit work so well? It simply reproduces nature's perpetual motion existing in every atom. The proof of this is the permanent magnet which maintains its constant magnetic field without an external input of energy. Suspend one ring magnet above another on a pencil (opposite polarities). Ask yourself, "What is providing the source of energy to continuously defy gravity and produce levitation?" It is the perpetual motion of the electrons from quantum mechanics and Pauli's exclusion principle, which has been called a force of nature by many physicists. Specifically, it is the *spinning motion of the electron* which is the major contributor to normal magnetism (over 95% of it), also called ferromagnetism. The orbiting motion of an electron, around the nucleus, contributes only to diamagnetism, *opposing* an external magnetic field [14].

High Temperature Superconductivity

In regards to a non-conventional approach to this subject, Ronald Bourgoin's patented process for drawing bismuth filaments that exhibit properties approaching superconductivity was the subject of his presentation at the First ISONCET, in 1981 (Pat. #4,325,795). The heat dissipation problem that currently exists for any space flight may become a worry of the past. Many other benefits, including healthful DC electricity, and new power-generation methods may evolve out of ambient temperature superconductivity (the name of Bourgoin's patent). The author filed an SBIR application back in 1986 to develop Mr. Bourgoin's invention in regards to ELF measurements but it was before room temperature superconductivity became a household word and the Navy rejected the application. Now, physicist Y.C. Lee (SUNY at Buffalo) and other scientists are publishing papers on the theoretical reasons for Angstrom-size bismuth filaments acting as "electron waveguides".

ATOMIC FUNDAMENTALS

People tend to forget that quantum mechanics *demands perpetual motion* of every atomic constituent. Otherwise, we could locate a particle to arbitrary accuracy. In that case, the universe would be completely different. It would be a completely mechanical universe where Descartes, Liebniz, and Newton would still be kings. The universe without perpetual motion would have *materialism* for a religion and every last action would be completely determined and predictable. Just a big machine behaving logically. In that universe, the elected president would probably be replaced by a small computer.

Energy Fluctuations

Proc. Intersoc. Energy Conver. Eng. Conference, 1991

However, we do live in a universe with perpetual motion. The only place where it theoretically does not exist is at absolute zero temperature, which is unattainable physically. The reader might say in opposition, "Oh, but most of this so-called 'perpetual motion'is just random fluctuations of atoms, etc. which could never be but to work." Well, we have to call your attention to Patent #4,004,210 (as well as #3,890,161) which is a *Reversible Thermoelectric Converter With Power Conversion of Energy Fluctuations* by Joseph C. Yater from Energy Unlimited, Inc. in Lincoln, MA. Refer to Yater's conference papers and his *Physical Review* articles [15]. To summarize: RANDOM ENERGY FLUCTUATIONS CAN BE PUT TO WORK AND THIS DOES NOT VIOLATE THE SECOND LAW OF THERMODYNAMICS.

Magnetic Motors

This discussion of the unusual behavior of electrons and permanent magnets raises the following question: Can magnetic motors be designed that use the magnetic energy of the electron's spin? Physicist and electrical engineer, Dr. Harold Aspden, believes there is a theoretical basis for such a belief, stating, "...if we ...operate with variable magnetic flux, then we can hope to draw energy from the intrinsic power sources that sustain the polarization of the ferromagnet" [16].

An inventor, named Howard Johnson, written up in the *New York Times* under the title, "Motor Run Solely by Magnets", also has a similar conviction [17]. His patented device, a magnetic motor, (#4,151,431) was demonstrated on the basis of linear models which showed unusual rapidity of motion propelling a specially designed, levitating magnet along the length of the track. An analytical, 34-page paper entitled, "The Permanent Magnet Motor" by Willam P. Harrison, Jr. of the Engineering Fundamentals Division of Virginia Polytechnic Institute and State University, Blacksburg, VA, was presented at several conferences and gained attention before the device was fully built. Today, Howard Johnson still is working on perfecting the model. Another inventor, Troy Reed from Reed Magnetic Motor Inc. recently has advertised a magnetic motor for sale [18]. With magnetic injectors, the inventor claims an indefinitely running motor, with moderate torque once the flywheel is started mechanically. A sampling of other patented magnetic motor inventions are referenced for further study [19].

Since magnets which are turned on or off (unshielded and then shielded) can produce large forces on conductors and other magnets, many inventors have persevered, even though a magnetic field is not conservative.

In other systems, such as thermal or electrical, we see that a temperature gradient or voltage gradient is enough to make extraction of energy feasible. An example of large-scale thermal gradient users is the 50 kW OTEC (ocean thermal energy conversion) plant off of the coast of Hawaii. We almost take for granted the pattern OTEC exhibits of using most of the 50 kW for generation (pumping water, etc.) with about 10 kW of "free energy" left over for distribution. Other systems to watch are the high efficiency heat pumps. These inventions with coefficients of performance (COP) of 7 or higher are the conventional free energy sources for the home which today produce heat and hot water, with more heat energy out than the small amount of electricity input used to run the efficient compressor [20]. COP, which is the ratio of output energy over input (with Watts converted to BTU's), is over-unity in most heat pumps. These are templates that can be followed for NCET generators.

Concerning magnetism, Maxwell's equations traditionally prohibit magnetic monopoles which could create a "magnetic field gradient". However, Prof. Shinichi Seiki, in his textbook, *The Principles of Ultra Relativity*, (10th edition), proposes use of the imaginary current usually left out of electromagnetic calculations [21]. He calls it "tachyon energy" (since tachyons also rely upon an imaginary term) and described in detail its use at the First ISONCET, held at the University of Toronto.

Proc. Intersoc. Energy Conver. Eng. Conference, 1991

The mobius coil, which has conductive strips on both sides of a dielectric and a single twist in the loop, is theoretically capable of conducting an imaginary current if we allow the magnetic intensity, H, to have an imaginary component [22]. Roles of electricity and magnetism become reversed in the derivation and the mobius coil is found to produce an equivalent imaginary magnetic "charge" or monopole.

As speculative as this may seem, Prof. Seiki has built up a textbook filled with his self-consistent theory. He describes many mobius generators which have anomalous electrical properties, such as steadily building up voltage over a period of weeks, or melting a stone with what looks like a high voltage arc produced from his low voltage circuit. In a recent correspondence, Prof. Seike mentions that he is now selling the basic components for researchers to duplicate his anti-gravity experiment involving a transistorized coil, detailed in his textbook. A price list for the four parts is available from him upon request.

Magnetic Generators

A simple, reversible magnetic motor, is the homopolar or "Faraday" generator. Discovered by Michael Faraday back in 1831, its unusual operation has eluded complete scientific explanation [23]. Interested readers are referred to the conference article, "The One-Piece Faraday Generator, Research Results" by this author.

Nuclear Battery

Analyzing the Hendershot and Moray devices, Paul Brown has taken them further by producing a 5 kW nuclear battery that lasts, based upon a beta emitter with a resonant circuit. Nucell, Inc. also has a small consumer battery which is the first NCET device ready for mass market [24]. He was a speaker at ITS 1990 [7].

PROPULSION

The push for new forms of propulsion methods has been intimately tied to new forms of energy generation. Most scientists and physicists recognize that in many cases a new form of propulsion often depends upon a new energy source that is converted into a propulsion source by electrical, mechanical, or thermodynamic means. The internal combustion engine is a good example. The chemical rocket is another example.

In fact, all propulsion methods follow this trend. Some are known to produce force without a readily apparent form of energy. Ion propulsion is an example. However, when we examine the application of ion passage through a magnetic field, we see that electric power is also available by the method of magnetohydrodynamics (MHD). Proving the above argument, recently, the Japanese developed an MHD-powered boat. To summarize: <u>EVERY PROPULSION SOURCE IS (OR ALSO DEPENDS ON) AN ENERGY SOURCE, NO MATTER WHICH IS DISCOVERED FIRST.</u>

Inertial Propulsion

Proc. Intersoc. Energy Conver. Eng. Conference, 1991

As an example of broader thinking on fields and forces, consider Mr. Robert Cook who has a forceful invention that works. Even though inventor Cook has not had any education beyond high school, he was able to think in cross-disciplinary terms. Looking at rotary systems in various machine shops while working, he wondered why centrifugal force couldn't be harnessed to do useful work. Further contemplation revealed the fact that the system resembled an electrical alternating current circuit. With that conclusion, he then wondered if it would be possible to "rectify" the alternating centrifugal force, by preventing some of the mass of a rotating body from making a full cycle.

Several years later and $100,000 poorer, he built the first "Cook Inertial Propulsion Engine" (CIP Engine). It was tested on ball bearings, on bicycle wheels, in a swimming pool, and on a perfectly flat and level granite table. His results may be seen in the **Proceedings of the First ISONCET** and on video. Due to its size and weight, it barely passed each test. However, he now has completed a second prototype that develops one pound (1 lb.) of thrust, powering a small boat. Among the 100 competing inertial propulsion patents listed in Cook's book, his is one of the simplest in principle. Ready for development, the CIP engine (pat. #4,238,968) method involves a variable, but even number of rotors, each of which are capable of producing a unidirectional force by keeping an exchangeable mass on one half of the rotation circle. The exchangeable develops centrifugal force during all of its motion but is restrained to only one half of a complete rotation by a unique method of oppositely spinning rotors. Cook describes the system as "reactionless" since he believes that he has outwitted Newton and can prove it.

A 30-page parametric study performed by Professional Engineer and Consultant Richard J. Rose, who has M.I.T. credentials and thirty years of experience, reveals an equation for maximum impulse with minimum rotor speed (p.30). The report also contains an average force equation for each rotor bank, graphs for average thrust versus RPM, and an acceleration equation for CIP operation in outer space, air, water, or land. One of the engineer's summarizing remarks reads, "For applications where loads must be lifted (and/or transported), the thrust F curves on

pages 11, 12, 17-19 can be used. In this case, the CIP engine is merely gimbaled into a vertical plane to produce lift" (p.30). This report is
available in its entirety from Robert Cook upon request [26]. Mr. Cook's
invention has been featured in many popular newspapers and journals
in the past few years as well as radio and TV appearances [27].

According to Mr. Rose, the CIP engine represents the highest efficiency of any propulsion method today, even in its current undeveloped
state. One significant point about the above invention is the fact that
the CIP engine can improve its efficiency even further by simply choos-
ing a good energy source. This is an example where a non-conventional energy source is a logical adjunct to a propulsion source.

Inventors not mentioned in Mr. Cook's book include Bruce DePalma,
who produced a propulsive force device using two force-precessed
gyroscopes that are rigidly attached to the same housing. The results
of Mr. DePalma's work was presented at the First ISONCET. Scott
Stracken also demonstrated a similar gyro at the Third ISONCET
which caused Dr. Aspden to start theorizing again [28]. Also worthy of
note is Roy Thornson, an inventor also at the Third ISONCET who
demonstrated an 8 lb. thrust.

Electrogravitics

An entirely different method of achieving propulsion is seen with John
Searle's approach, whose unusual high voltage method for achieving
propulsion is best described in a book entitled, *Ether Technology* [29].
It is well illustrated and also summarizes Mr. T. Townsend Brown's
less effective electrostatic method for propulsion, also called
"electrogravitics". It is worth noting that in the early 50's,
electrogravitics was a classified government project, as proven by the
text released recently by the Freedom of Information Act. Dr. Paul

Proc. Intersoc. Energy Conver. Eng. Conference, 1991

LaViolette recently informed me that one copy still exists at the Library of Congress.

Kinetobaric Propulsion

During 1981, the author conducted an initial investigation into one of the most unusual non- conventional propulsion inventors of the First ISONCET mentioned above. An example of a first-class overseas inventor is Rudolph Zinsser of W. Germany. He has produced a very viable and demonstrable method for propulsion that is unequalled anywhere else in the world. Mr. Zinsser's invention relies upon his patented signal generator (Pat. #4,085,384) to effect a force in a specially-designed capacitor using a water dielectric. He has been able to produce several dynes of force with an apparatus only about 10 cc in volume. The unique property of Mr. Zinsser's invention is the storage capacity of the system that is irradiated with the patented impulse signal. The effect is most closely described as producing a "local gravitational anisotropy". During many private communications, Mr. Zinsser provided much stripchart data substantiating the constant force or thrust given to a target, lasting for several minutes after the signal was terminated. He has referred to this as a "kinetobaric" effect and works with Dr. Peschka from a nearby university who has developed a theory to explain the effect. The output has been measured at 6 Newton-sec per Watt-sec, or about 25 pounds of thrust for 100 Watts of electrical energy input. The potential for upscaling this invention is enormous. In 1981, the author was actively involved in securing funding sources for Mr. Zinsser's invention. Mr. Zinsser was a guest speaker at the First and Second ISONCETs. At this time it appears that Mr. Zinsser has found funding in West Germany for he continues to perfect his invention to this day.

CONCLUSION

Why do scientists have so much difficulty with cross- disciplinary insights? Perhaps it is due to their narrow education which did not bridge the disciplines while teaching the basic physical principles as "Unified Technical Concepts" (UTC) does for physics [25]. In this Freshman college course, the similarities between electrical, mechanical, fluidic, and thermal systems are emphasized. It could be called a "Systems Theory" approach to physics. There are "force-like" concepts and "distance-like" concepts in all four systems mentioned above. The beauty of this teaching method, from my experience, is that it helps the student to remember formulae which are based upon *more fundamental, analogous principles.* As more unified physics is taught, future scientists will evolve who produce new methodologies that will broaden the scope of energy research and development, as a direct result of their comprehensive, unified conceptual insight.

With the enormous demands of the Space Shuttle payloads, an available replacement propulsion method is easily 10 to 20 years in the future. Dr. Gerard O'Neill's solution (author of *Space Colonies*) is to use the resources of the moon and asteroids to avoid trips to and from the earth. However, the voltage, current and power density limitations that presently are insurmountable for low earth orbit have to be considered before sizable interplanetary spacecraft are produced. For further study of the popular non-conventional energy inventions, the reader is referred to *The Manual of Free Energy Devices and Systems* by Donald Kelly [30].

Integrity Research Institute is committed to the NCET revolution and welcomes your inquiries and support.

Proc. Intersoc. Energy Conver. Eng. Conference, 1991

REFERENCES

[1] Valone, T.F., "Exploitation of Gravity Field Energy: Hannover 1980", *Energy Unlimited*, Vol.1 No.9, 1981, p13.

[2] Neiper, Hans, *Conversion of Gravity Field Energy*, Ilmer verlag, Greifswalder Strasse 2, D-3000 Hannover 61, W. Germany, 1981.

[3] Neiper, Hans, *Dr. Nieper's Revolution in Technology, Medicine and Society*, MIT, POB 4328, Huntsville, AL 35815, 1985.

[4] *First, Second and Third International Symposia on Non-Conventional Energy*, proceedings available from Planetary Association for Clean Energy (PACE), 100 Bronson Ave. Suite #1001, Ottawa, Ontario. Also published by cosponsor: Cadake Industries, POB 1866, Clayton, GA 30525. (Request Cadake non-conventional energy book catalog. It's one of the best.)

[5] George Hathaway, P.Eng., 39 Kendal Ave., Toronto, Ontario M5R 1L5, consultant for NCET investors and inventors, is available at 416-923-8586.

[6] U.S. Psychotronics Association (USPA), 2141 West Agatite Ave., Chicago, IL 60625. Yearly conference video and audio tapes available upon request. The *Journal of the USPA* also available by subscription.

[7] International Tesla Society, affiliated with the IEEE holds conferences every two years (even years). Proceedings, videos and audio tapes available from High Energy Enterprises, POB 5636, Security, CO 80931

[8] Meeting of the Minds Conferences, sponsored by AZ Industries, video tapes available. Many papers from the conferences also are

found in *Magnets in Your Future* magazine, published by AZ Industries, Hardy, Arkansas.

[9] *Congress For Free Energy, 1989 Proceedings*, ISBN 3- 9520025-1-8, just published. Available from SAFE, Postfach 402, CH-8840, Einsiedeln, Switzerland.

[10] T. Henry Moray, *The Sea of Energy*, Cosray Institute, Salt Lake City, Utah, Available from [7].

[11] Energy Unlimited Publications, P.O. Box 413, Wilton, Iowa 52778 (ask about back copies of magazine)

[12] Prigogine and Stenger, *Order Out of Chaos*, MIT Press, p. 308.

[13] These figures are from the textbook for the course, *Space Power Technology*, Gilmour, Hyder and Rose, SUNY at Buffalo, ECE Dept., Buffalo, NY 14260, 1985.

[14] Chikazumi & Charap, *Physics of Magnetism*, J.Wiley & Sons, NY, 1964, p.41

[15] Joseph Yater, "Energy Fluctuations", *Physical Review A*, Vol.20, No.4, Oct. 1979 P.1614. (A second followup article was published shortly after this in response to a reader's comments on this one.) Also see, Second International Conference on Thermoelectric Energy Conversion at the U. of Texas at Arlington, 3-22-78.

[16] Dr. Harold Aspden, "The Vacuum as Our Future Energy Source", *Magnets in Your Future*, August, 1988, p.15.

[17] "Motor Run Solely by Magnets", Stacy Jones, *N.Y. Times*, 4-28-79, p.32. See also: *Science and Mechanics*, "Amazing New Motor Powered only be Permanent Magnets", Spring, 1980, p.45 and *Popular Science*, "Magnetic Wankel for Electric Cars", June, 1979, p.80.

Proc. Intersoc. Energy Conver. Eng. Conference, 1991

[18] Troy Reed, Reed Magnetic Motor Inc., POB 700395, Tulsa, OK 74170

[19] Other patented magnetic motor inventions include, "Energy Conversion System", J.W. Putt, #3,992,132; "Magnetic Piston Machine", R.F. Stahovic, #4,207,773; "Magnetic Motor", S.Kuroki, #4,305,024; and "Permanent Magnet Motor Conversion Device", J.W. Ecklin, #3,879,622 and "Biased Unitized Motor Alternator with Stationary Armature and Field", J.W. Ecklin, #4,567,407, to name a few. Ecklin's invention has received a lot of attention.

[20] Dennis Lee's United Comm. Services presentation from Meeting of the Minds Conf., AZ Industries, 1988.

[21] Shinichi Seiki, *The Principles of Ultra Relativity* published by the Space Research Institute, Box 33, Uwajima (798), Japan. Phone: 0895-24-0225.

[22] Such as the AEC-NASA Tech. Brief, #68-10267, July, 1968 "Mobius Resistor is Noninductive and Nonreactive"

[23] Thomas Valone, *The One-Piece Faraday Generator,* 4th ed., 125 pgs., pub. by Integrity Research Institute.

[24] Paul Brown, Nucell, Inc., 12725 SW 66th Ave., Suite 102, Portland, OR 97223. 503-624-8586.

[25] *UTC-Physics for Technicians,* Center for Occupational R&D, Waco, TX, 1983. Another good text in this area is *Unified Concepts in Applied Physics* by Dierauf and Court, Prentice-Hall, 1979.

[26] *Death of Rocketry* by Robert Cook, Cook Pub., 16259 Llanada, Victorville, CA 92392

[27] *Vallejo Independent Press,* 2-21-81, Odessa American, 9-29-85, and the Times-Herald

4

The One-Piece Faraday Generator:
Research Results

THE ONE-PIECE FARADAY GENERATOR: RESEARCH RESULTS

Thomas Valone, M.A., P.E.

Proceedings of the Intersociety Energy Conversion Engineering Conference

1991

Integrity Research Institute
1377 K Street NW, Suite 204
Washington, DC 20005

THE ONE-PIECE FARADAY GENERATOR:
RESEARCH RESULTS

Thomas Valone, M.A., P.E.

Integrity Research Institute

1377 K Street NW, Suite 204

Washington, DC 20005

ABSTRACT

Faraday's disk experiments of 1831 have significance for a variety of reasons. From rail guns to Tokomaks to the origin of the earth's magnetic field, the Faraday generator has played a key role in present day science. Also called a homopolar, unipolar, or acyclic generator, it is the only one to produce electricity without commutation. Faraday's one-piece style of co-rotating the cylindrical magnet with the conducting disk is considered to be an unusual configuration and has eluded complete scientific explanation. To this day, prominent scientists can be found who believe it will not work since they operate with the flux line conceptualization. However, its importance is found in the connection to the earth's magnetic field, which evolves from a one-piece Faraday generator. A laboratory model is used to investigate the presence of back torque or armature reaction with the generation of electricity. For the first time, 1) the back torque of a one-piece homopolar generator has been measured, 2) the classification of the homopolar generator ary in space, whether or not the magnet rotates with the disk, creates the electromotive force (emf) measured from center to outer edge of the Faraday disk. A detailed discussion of the field rotation paradox may be found in my book, **The One-Piece Faraday Generator, Theory and Experiment** [2]. The simplest explanation for the operation of the HPG or the OPFG is the application of the Lorentz force on arbitrary radial segments as they pass through the magnetic field, thus explaining the force on the conduction layer electrons. Further molecular lattice effects of the motoring effect of back torque can be understood in terms of the Hall effect and the force on positive ions [10].

As the current is generated by the emf, a negative spiraling effect is seen as the disk rotation leads the radial current around the disk on its way across it. A further experiment is possible with a radially-segmented disk, to eliminate these eddy losses which tend to demagnetize the field. An alternative which Tesla proposed, is a *spirally-segmented disk*, which becomes a self-exciting Faraday generator (SEFG) [8], countering demagnetizing effects. Tesla's suggestion eliminates the problem of the standard current flow creating a partial eddy current of its own. A wise note of Tesla's is to optimize the design of the spiral with the operating speed, thus preventing any negative effect from excessive speed. Integrity Institute has plans to use computer modeling for a spirally-segmented OPFG.

THE EARTH'S FARADAY DISK DYNAMO

Self-Sustaining Vs. Self-Exciting

One of the major reasons for interest in the OPFG is that the earth itself functions internally as a large OPFG. Moreover, the earth's OPFG

Proc. Intersoc. Energy Conver. Eng. Confer., 1991

as a regulated voltage source has been experimentally determined, and 3) an effect, involving the lack of measurable voltage in the rotating frame, has been verified with a specially designed LED voltmeter. A back torque value of 0.17 N-m for a 25 Watt generator was obtained, in agreement with theory.

INTRODUCTION

The general homopolar generator (HPG) is one in which a disk or a drum is rotated *adjacent* to a magnet of the same size and shape. It has been suggested by DePalma that the one-piece Faraday generator (OPFG) may have the unusual possibility of the absence of back torque [1]. Subsequently, the author [2], Trombly [3], and Wilhelm [4] began three independent experiments to replicate DePalma's results. Only one of us claim success in that endeavor, while Wilhelm and the author experienced back torque which compensated for the generated power in most cases. All three of the above scientists used liquid metal brushes in their experiments (Trombly-NaK; Wilhelm-Hg; Valone-low temperature solder) to reduce contact resistance. It is agreed that Trombly's sodium-potassium, having the viscosity of water, was superior to the other two. A major problem affecting all liquid metal brushes is the MHD instability caused by electrical conduction and motion in the presence of the magnetic field. None of us have calculated the measurable effect due to MHD that may have contributed to our results but they are expected to be negligible [5]. Referred to as an electromagnetic pumping force [6], the liquid metal becomes turbulent when the Reynolds number exceeds 2000. Eddy current and MHD losses then occur [7].

Eddy currents in the solid conducting disk are not a contributing factor to losses since there is no changing magnetic field. However, the motion of the conductor through the magnetic field, which remains station-

Proc. Intersoc. Energy Conver. Eng. Confer., 1991

is self-sustaining. "The crucial question is how the core liquid flows to act as a dynamo. Also a self-sustaining dynamo does not require a constant supply of magnetic field, it does require a constant supply of mechanical energy to keep the conducting material moving. In the case of the earth's core this means not only that the metallic fluid must flow in the right manner but also that some energy source must sustain the flow" [9]. Helical convection patterns called "rollers" created from conducting liquid metal are the best explanation of the mysterious secret of the earth's self-sustaining OPFG (SSOPFG).

In regards to the back torque of the earth's SSOPFG, Busse, Roberts, Lowes, and Wilkinson of the University of Newcastle upon Tyne are working on mechanical models of the earth's core to explain the changes in the fluid's speed and direction when the magnetic forces are large. A slightly different model that is being tested as well is the self-exciting OPFG (SEOPFG) which requires a spirally-segmented disk and/or external current-carrying coils as Tesla suggested. Since he noted that the armature current tends to demagnetize the field, in a normal solid disk configuration, Tesla felt that the subdivision of the disk would be an enhancement. In regards to these beneficial eddy currents, he writes, *"The current, once started, may then be sufficient to maintain itself and even increase in strength, and then we have the case of Sir William Thomson's 'current accumulator'"* [8].

A laboratory SEOPFG has been built by the Lowes and Wilkinson team [11]. Using metal rollers to simulate the earth's cylindrical eddy currents, the team found some interesting results after beginning with a few viscosity problems. "...a more efficient geometry was found, so efficient that the dynamo would self-excite in a completely homogeneous state (i.e. with *no insulation*) at a much lower rotor speed than was believed possible" [12]. Upon achieving this breakthrough, their next goal is to look into the stability of the dynamo mechanism, hoping to observe reversals of its magnetic field.

An illustration of a self-excited Faraday generator (SEFG) is shown in Fig. 1, where the implication is that the model is a portion of the earth's

SEOPFG [13]. The concept of the SEFG is used in some applications when an electromagnet is desired [14]. It is possible to use dual SEFGs, each exciting the other, by cross-connecting the windings. Furthermore, by creating two independent windings on each machine, with the fluxes adding on one and subtracting on the other, one can obtain two-phase alternating current [15].The AC power output of the dual SEFGs has self-limited oscillation of the magnetic field polarities as well! Being a high current, low voltage device, the FG expands its range of applications with this AC improvement.

Through further study of the SEOPFG and the SSOPFG, we hope to strive toward Tesla's prediction of an energy accumulator or at least to approach the earth's amazing SSOPFG, (which is made entirely from molten metal). Whether the SEOPFG may become the free energy generator of the future, solving home electrical power needs, as Trombly and DePalma believe, remains to be seen. (More information about free energy can be found in the other article by this author, "Non-Conventional Energy and Propulsion Methods" published concurrently in this **IECEC Proceedings**.)

One very pleasing discovery, that has not been found in the literature, was made with an early model of the OPFG. The OPFG, even the presence of a generated current, does not diminish or depress the emf, as normal voltage supplies and batteries do when loaded. The HPG and OPFG behave exactly as a *regulated voltage source* does. When tested at a fixed speed, the voltage remains the same no matter how much current is drawn from the generator [19].

Effects of the Earth's Faraday Generator

Historically, Faraday looked for the effects of the earth's OPFG, thinking that the electromotive forces (emfs) could be measured on the rotat-

Proc. Intersoc. Energy Conver. Eng. Confer., 1991

ing disk. He tried to measure these emfs in rivers and streams [16]. Other scientists have committed a similar mistake, notably Corson [17], not knowing that such emfs are equally cancelled within the rotating frame by a self-created electrostatic field oppositely directed by the charge displacement [18]. Because **B** is not changing in time, **curl E** = 0. Therefore, the electric field that is created must be irrotational, i.e. electrostatic. (A lab test of the electrically neutral environment in the rotating frame is summarized in the Laboratory section.)

Though the emf of the earth's SSOPFG is not measureable on the surface of the rotating earth, some scientists believe the emf effect is most noticeable in the aurora borealis [20]. In fact a few have calculated the voltage that should be measurable from the pole to the equator in the magnetosphere [21]. Furthermore, some have even attributed the same HPG effects to the electromagnetic fields of stars [22].

RELATIVITY AND THE FARADAY DISK

Though space limitations prohibit a full theoretical analysis, most treatments of the HPG and the OPFG reviewed start with the Lorentz force to calculate the **E** field (**E** = **v** X **B**) and measurable voltage (emf). However, Becker notes in regards to the HPG, "we need the concept of the Lorentz force which is foreign to the Maxwell theory but is derivable by relativity theory" [23]. Then the nonvanishing divergence of the HPG effective electric field (E=Bwr, where w is the angular velocity) leads to a volume charge density for the generator, which correlates with the electrostatic field derived earlier. Library research revealed that special relativity can be used to describe the HPG and the OPFG, through use of the polarization/magnetization vectors of electromagnetism which are oppositely paired [24]. However, since the rotating disk is truly a non-inertial reference frame, general relativity must strict-

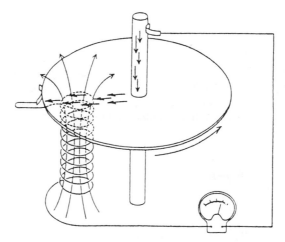

Figure 1

RELATIVITY COMPARISONS	
Rectilinear Motion	Circular Motion
· No voltage developed when bar and meter move together	· Voltage not developed when disk and meter move together, but electric field is generated
· No difference between motion of observer and charge: $M_i = V \times P_i$	· Difference between rotating charged sphere or rotating observer (Schiff, 1939) B Field vs. no field · Ring currents developed causing magnetic field for sphere rotation
· No absolute motion detectable	· Absolute Rotation measured (wrt inertial frame) Sagnac, Mannov; see Mannov, Foundations of Physics Vol. 8, 1978 p.137
· Special Relativity applies	· Special relativity doesn't apply
· No volume charge by special relativity transformation laws	· Volume charge: $E = V \times B$ $D = c_i E$ $\rho = \nabla \cdot D$ $\rho = -2\epsilon_i \omega B$
· No forces for uniform, constant velocity	· Centrifugal and coriolis forces generated

Figure 2

ly be used [25]. Fortunately, since the curvature tensor vanishes, space is flat and the Lorentz metric is applicable, though various authors derive slightly different results using general relativity [26]. A table has been assembled to summarize the interesting relativistic facts that various authors have contributed to this subject (Fig. 2) [27].

LABORATORY RESULTS

A One-Piece Faraday Generator With Liquid

Metal Brush

With three preliminary models, a fourth working prototype of the OPFG was fabricated using 8" ceramic magnets (four on each side of disk) and a 1/2" thick copper disk (Fig. 3). A General Electric current shunt (50 mV @ 2500 A) was used for accurate current measurement and a DC motor with 12 V battery power. Between Trombly, DePalma, Wilhelm and the author, however, this OPFG had the highest internal resistance, (which alone limits the theoretical maximum power output). Using an Electronics Limited Milliohmeter (Model 47A), it was measured at 230 microhms (+/- 10) with the current shunt disconnected. The GE current shunt contributed 20.0 microhms to the circuit while the contact resistance of the milliohmeter leads was about 30 microhms. Noting the contact resistance of copper brushes, the danger of mercury, and the wetting problem of mercury, it was decided to fabricate a circular trough with AC heaters for a low temperature solder (Wood's Metal) brush. Through one year of trial and error, it was discovered that the brush of an HPG has to be located as close to the disk as possible to obtain the maximum emf and current. Therefore, a

Proc. Intersoc. Energy Conver. Eng. Confer., 1991

special circular flange was designed (see Fig. 4) to pass through the circular trough filled with hot, melted solder.

Without the internal resistance of the liquid metal brush, the internal resistance of the copper disk and 1" brass shaft was measured to be about 0.1 microhm. Woodson and Melcher show that the maximum current can be calculated using the open circuit voltage, which yields about 500,000 Amps, at 1000 RPM. The maximum power that could therefore theoretically be delivered at that speed is about 30 kW [28]. While impressive, this shows the vital importance of a very low resistance brush system for high power output HPGs. For example, with the help of superconducting magnets, Northern Engineering Industries in England has designed an HPG capable of 1300 megawatts of continuous output [29].

Testing for Output and Back Torque

The 8" ring magnets used were the Ferrimag 5 #MF-51239 which have a 4" hole in the center, where the flux actually reverses. Though the center flux is of much lower intensity than the rest of the magnet area, the effect forced the the average magnetic flux density (B) to become a calculated quantity. Using the standard V=wBR/2, where R is the difference of the *squares* of the inner and outer radii, we solved for the average flux density experienced by the OPFG during open circuit emf production. For six trials, the average *B = 0.163 Tesla*.

With B determined, the torque delivered to the generator by the DC motor could be calculated for each trial, using P=wT. The GE motor was a 24 V, 20 A, 3400 RPM, 1/2 Hp motor. However, since the efficiency curve could not be obtained from the dealer nor from GE, we were forced to assume 100% efficiency for all power calculations. Since the load to the motor varied less than 10% between open and

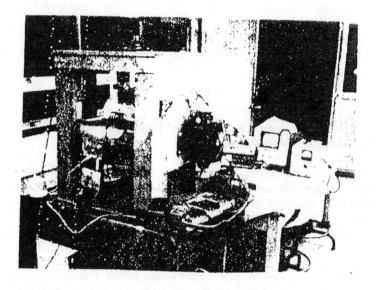

Figure 3

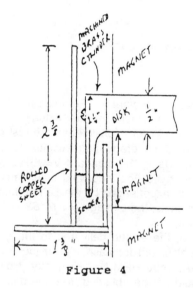

Figure 4

closed circuit generator operation, the efficiency variation was expected to be minimal, though the efficiency itself was most likely less than 100%. (See Error Discussion section for more information.)

In one sample trial, with stripchart results shown in Fig. 5, at 1000 RPM, the open circuit voltage was measured to be 65 mV (bottom graph) while the drive motor power consumption was 249 watts, which yielded a torque calculation of 2.37 N-m. At the same speed, the short circuit current was measured to be 380 Amps (top graph), with the drive motor power consumption going up to 266 watts, yielding a torque calculation of 2.54 N-m. Taking the difference of the two torques, *we find 0.16 N-m extra torque* was needed to drive the generator during power output. Therefore, this is an indirect measure of the back torque of the generator. A digital tachometer was used to verify the speed, which was maintained at 16.7 Hz (1000 RPM).

In spite of this back torque, and the accuracy of the voltage and current measurements, the drive motor power consumption increased by only 17 watts while the generated power was 24 watts, with an estimated error of +/- 2 watts. This anomalous power output, often referred to as "free energy" [30], cannot be explained readily. The clear current output line of this trial shows good continuous conduction without the turbulence that plagued many further tests as the solder overheated. However, after teaching physics for several years and emphasizing error calculations, it becomes apparent that the relative errors in the three place accuracy of the power calculation become a source of the problem when the *subtraction* is made to obtain the difference in the motor power demand. The relative error is calculated to be at best +/- 6 watts and at worst, +/- 12 watts. *Therefore, the 17 watts is really accurate to only one digit.*

A second sample trial is shown where conduction through the liquid metal was hampered slightly and turbulence is apparent in the current output graph. Here the generator was operated at 600 RPM, with 45 mV and 240 A produced. Drive motor power consumption was 139 W open circuit and 150 W closed circuit, yielding a difference of 11 W.

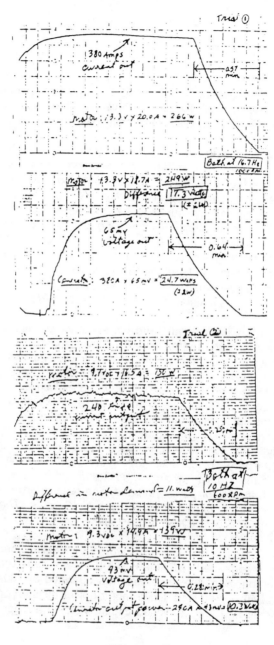

Figure 5

Proc. Intersoc. Energy Conver. Eng. Confer., 1991

Torque to the generator was 2.39 N-m open circuit and 2.21 N-m closed circuit, yielding 0.18 N-m difference which is again within 10% of the expected range, in spite of the relative error discussion above, *which applies to torque measurement/calculations as well.* In this case however, the power difference of 11 watts was almost exactly equal to the generated power of the OPFG. It may be possible that a loss was created in the erratic conduction through the brush, increasing the resistance of the brush and therefore, decreasing the current output. This would cause the generated power to drop as well. Fresh solder should probably be used for each trial since it oxidizes.

Comparing with the theoretical calculation of back torque (using $\mathbf{T} = \mathbf{J} \times \mathbf{B}$) from the point of view of the torque generated from the passage of generated current through the magnetic flux of the system $T=BRI/2$ (with the same definition of R as above), yields 0.164 N-m which is in close agreement with the torque difference method above. This calculation shows that back torque is really the same as a homopolar motor (HPM) effect, analogous to the back emf in motors. Researchers therefore try to maximize HPG effect while minimizing the HPM effect. Utilizing a 1) closed path magnetic field, as Trombly and Kahn, 2) a low reluctance disk (iron or steel), and 3) a spirally-segmented disk will all contribute to changing the balance of the unaltered or "natural" HPG and OPFG.

Testing Rotating Frame Voltage

A test of the relativistic effect of a neutral electric environment in the rotating frame of the OPFG disk was performed, *even in the presence of generated current.* Using a modification of a previously designed voltage regulator which has an internal voltage reference [32], an LED voltmeter was placed on the rotating OPFG to look for the presence of any voltage surpassing an arbitrary 15 mV threshold. Tested in the laboratory rest frame, with the solder-soaked leads sliding on the shaft and periphery, the LED turned off almost immediately as the OPFG started turning, generating over 100 mV. Designing the LED voltmeter to *turn off* as it measured the voltage was of great value for the high speed rotor motion. Since the LED circuit becomes part of the generator as it is rotating, it has been suggested that it cannot function because it is a conductor itself. However, since the LED voltmeter is a 9 volt system, *and stays lit,* the electrons in the low voltage emf environment are not overpowered by the effective electric field within the rotating frame. The small circuit, about the size of the 9 volt battery that was used, is diagrammed in Fig. 6.

ERROR DISCUSSION

Besides the previously mentioned relative error discussion in third digit precision that shows up in subtraction of measured values, the perplexing counter argument to the anomalous Trial #1 centers on the DC motor's efficiency. If the efficiency is assumed to be about 90%, for example, instead of 100%, then we have the interesting problem where the transfer function of the motor is a constant 0.9 multiplied by the torque or the battery power delivered. This means that the values recorded for the "torque to generator" would be 10% lower and the *difference*, being the back torque of interest, would be 10% less as well. This creates an even greater advantage to the free energy advocates, who would see the OPFG becoming over-unity in this example.

For future experiments, it is important to have efficiency curves for the motor in use and to measure to four-place accuracy wherever possible. The fascinating discoveries of the OPFG in the lab, the connection to the earth's own core activity, and the prediction by Nikola Tesla make the interest in the OPFG justifiably increase year after year.

Proc. Intersoc. Energy Conver. Eng. Confer., 1991

REFERENCES

[1] Bruce DePalma, **DePalma Institute Report,** No. 1, 1978. (The DePalma Institute is located in Santa Barbara, CA)

[2] Thomas Valone, **The One-Piece Faraday Generator, Theory and Experiment,** pub. by Integrity Research Institute, 4th ed., 125 pgs. 1988. Also see "The One-Piece Homopolar Generator", **Proceedings of the First and Second International Symposium on Non-Conventional Energy Technology**, 1981, 1983. Cadake Industries Pub., Clayton, GA. Also see "The Homopolar Generator: Tesla's Contribution", **Proceedings of the International Tesla Society Conference,** 1986, Colorado Springs, CO. Preliminary lecture: The Symposium on Energy Technology, Hanover, W.Germany, 1980.

[3] Trombly and Kahn, International Patent #WO 82/02126 Adam Trombly has not published results of his experiments with the NaK OPFG, performed under the auspices of the Acme Research Corp., but presently can be contacted through the Earth First Foundation, Evergreen, CO.

[4] Timothy Wilhelm, **Stelle Letter**, Vol.15, No.9, 10/80.

[5] Hong and Wilhelm, **J. App. Phys.**, 47, 906, 76.

[6] A.K. Das Gupta, **I.E. (I) Journal E-L**, 50, 48, 69

[7] D.L. Lewis, **J. Sci. & Tech.**, 38, 47, 1971

[8] Nikola Tesla, "Notes on a Unipolar Dynamo", **Electrical Engineer,** Sept. 2, 1891, p.258.

[9] Carrigan & Gubbins, "The Source of the Earth's Magnetic Field", **Scientific American**, Feb., 1979, p. 118

[10] W.R. McKinnon, et al., "Origin of the Force on a Current-Carrying Wire in a Magnetic Field", **Amer. J. Phys.**, 49(5), May, 1981, p.493

[11] Hindmarsh,Lowes, Roberts, and Runcorn, **Magnetism and the Cosmos**, American Elsevier Pub. Co., 1965

[12] Hindmarsh, et al., p. 124.

[13] Carrigan and Gubbins, "The Source of the Earth's Magnetic Field", **Scientific American**, Feb., 1979, p.122

[14] Sears Patent #3,185,877.

[15] M. Zahn, **Electromagnetic Field Theory**, J. Wiley, NY, p. 423

[16] Michael Faraday, **Experimental Researches in Electricity**, reprinted 1965

[17] Lorrain and Corson, **Electromagnetic Fields and Waves**, W.H. Freeman & Co., San Francisco, 1970, p.226

[18] Panofsky and Phillips, **Classical Electricity and Magnetism**, Addison Wesley, Reading, MA, p.149 as well as Culwick, **Electromagnetism and Relativity,** J.Wiley & Sons, NY, 1962, p.143.

[19] Surprisingly, this was verified with only a 1" OPFG using Samarian-Cobalt magnets. As the resistive load varied, the emf did not change.

[20] Pogg, **Ann.**, 1852, p.357

Proc. Intersoc. Energy Conver. Eng. Confer., 1991

[21] F.J Lowes, "The Earth as a Unipolar Generator", **J. Phys. D: App. Phys.**, Vol. 11, 1978, p.765

[22] Leverett Davis, Jr., "Stellar Electromagnetic Fields", **Physical Review**, Vol.22, No.7, Oct.1,1947,p.632. Also see: E.N. Parker, "Magnetic Fields in the Cosmos", **Scientific American**, Aug. 1983, p.44

[23] Richard Becker, **Electromagnetic Fields and Interactions,** Blaisdell Pub., NY, p.378

[24] Panofsky and Phillips, **Classical Electricity and Magnetism,** Addison Wesley, Reading, MA, p.338

[25] Adler, Bazin, Schiffer, **Introduction to General Relativity**, McGraw Hill, 1975, p.257

[26] Webster, "Schiff's Charges and Currents in Rotating Matter", **American Journal of** Physics, 31, 590, 1963 and also Ise and Uretsky, "Vacuum Electrodynamics on a Merry-Go-Round", **American Journal of Physics**, 26, 4341, 1958

[27] Sagnac, **Compt. Rend.** 157, 708, 1410 1913 and Schiff, **Proc. Nat. Acad. Sci.**, 25, 391, 1939.

[28] Woodson and Melcher, **Electromechanical Dynamics,** J. Wiley, NY, p.288

[29] **Cryogenics**, Sept. 1982, p.435

[30] Thomas Valone, "Non-Conventional Energy and Propulsion Methods", **Proceedings IECEC**, 1991

[31] Valone & Shih, "Protected Regulator Has Lowest Dropout Voltage", **Electronics**, April 24, 1980, p.130

5

Macroscopic Vacuum Polarization

MACROSCOPIC VACUUM POLARIZATION

by

Moray B. King, author of *Tapping the Zero Point Energy*

ABSTRACT

Nikola Tesla and T. Henry Moray claimed inventions that apparently absorbed anomalously large amounts of environmental radiant energy. These inventions may have utilized a macroscopic vacuum polarization zero-point energy coherence associated with ion-acoustic oscillations in a plasma.

INTRODUCTION

It is generally taught to engineers that Maxwell's equations constitute a complete theoretical basis for all macroscopic electrodynamics at the engineering level. On the other hand it is also taught that other effects (e.g., pair production, vacuum polarization, zero-point fluctuations) can occur that are not predicted by Maxwell's equations, but these are at the "quantum level." It is tacitly implied that quantum vacuum effects do not relate to "engineered" systems. Yet in reality there is only one electrodynamics and to totally engineer it requires an understanding at its basis.

Moray B. King

Quantum electrodynamics shows that the basis of all electrical phenomena is the vacuum, where tremendous fluctuations of electrical field energy occur; this energy is called the zero-point energy,[1-4] and by some consideration is the modern term for the ether. Tesla[63] and later Moray[60] believed their devices interacted directly with the ether. Today's physics recognizes that matter interacts with the vacuum zero-point energy.[1] The term commonly used to describe the interaction of charged particles with the vacuum is "vacuum polarization." Whereas, many investigators use the term only to mean pair production, here it shall include all the states of the vacuum from effects in the low field linear regime (where Maxwell's equations apply), through the nonlinear regime on the threshold of pair production, to the "bifurcation catastrophe"[5] regime where pair production occurs.

In this paper, literature is identified that shows that the zero-point energy is necessary in electrodynamics and the manner in which it is needed to explain the radiation of a uniformly accelerated charge. References are also cited that demonstrate that the various elementary particles interact differently with the zero-point vacuum fluctuations. It is suggested that as a result of this interaction, ions may have different radiation characteristics than conduction electrons. This motivates the hypothesis that the ion-acoustic mode of a plasma may produce a propagating, coherent, macroscopic vacuum polarization. A qualitative vacuum polarization model is proposed to explain why conduction electrons would not readily detect this type of radiant energy. It is suggested that the ion-acoustic oscillations in Tesla's and Moray's devices absorbed longitudinal, vacuum polarization displacement currents, and that these inventors actually did discover a novel form of environmental radiant energy.

UNIFORM ACCELERATION OF CHARGE

Any complete theory of electromagnetism must include the zero-point energy, for ignoring it leads to contradictions

Macroscopic Vacuum Polarization

and paradoxes at a very fundamental level. One difficulty in electrodynamics is known as the equivalence paradox. Here a uniformly accelerated charge is recognized to radiate. However, a charge suspended at rest in a uniform gravitational field does not. According to general relativity, a uniformly accelerating system in free space should be equivalent to one at rest in a uniform gravitational field. Thus, in this case the principle of equivalence seems to be violated. This problem has been discussed in the literature at the classical level without adequate resolution. For example, Rolrhich,[6] Atwater,[7] and Ginzburg[8] conclude that radiation is a function of the acceleration of the observer in relation to the source charge. But as Ginzburg asks, what are photons, and what propagates at the velocity of light if it can be made to appear or disappear depending on the acceleration of the observer? Boulware[9] similarly suggests that "the way out of the paradox is to deny that the concept of radiation is the same in the accelerated and unaccelerated frames."[9] This interpretation likewise throws out the independent existence of light by linking it to the motion of the observer.

C. M. Dewitt[10] and B. S. Dewitt[11] acknowledge the violation of equivalence but state that electric charge is an "unfair," i.e., scientifically invalid, test of the principle of equivalence since a real gravitational field is only uniform locally, but a charge's field persists to infinity. B. S. Dewitt[11] also states that spinning neutral bodies also deviate from geodesic motion and they too are "unfair" tests of the principle of equivalence since angular momentum is a manifestation of a "nonlocal" phenomenon. It appears that either one must admit some of the laws of physics disobey the principle of the equivalence or that light cannot exist as an independent entity. The equivalence paradox has not yet been adequately resolved at the classical level.

An even more basic problem appears in classical electrodynamics regarding the uniform acceleration of charge. It is generally accepted that the radiated power is proportional to the square of the acceleration as computed by the Larmor formula:[12]

$$p = \frac{2}{3} \frac{e^2 a^2}{c^3}$$ (1)

Yet the radiation reaction friction force experienced by the charge is proportional to the first time derivative of the acceleration:[12]

$$F = \frac{2}{3} \frac{e^2}{c^3} \frac{da}{dt}$$ (2)

For uniform acceleration this derivative is zero while the acceleration is not. The particle radiates but does not lose kinetic energy. Where does the radiation energy come from? Fulton,[13] Ginzburg,[8] and Boulware[9] conclude it has to come from the charge particle's source field. But Pauli,[14] Vasudevam,[15] and Luiz[16] come to the opposite conclusion, stating, "...we cannot accept the assumption of radiation from the charge because otherwise the internal energy of the particle should be exhausted."[16] Luiz further argues, "The law of action and reaction is a fundamental law in physics and if the radiation reaction is zero, certainly there is no radiation."[16] Ginzburg[8] points out that this fundamental problem manifests itself in attempts to match theory with experiments measuring synchrotron radiation. Surprisingly, Leibovitz[12] concludes that Maxwell's equations are incompatible with uniform accelerating motion! Electrodynamics clearly suffers serious problems at the classical level where the zero-point vacuum fluctuations are ignored.

ZERO-POINT ENERGY

When the formalism for the zero-point energy is introduced, some of these issues may be better understood and resolved. Callen[18] demonstrates that the vacuum fluctuations manifest themselves even at a classical level: "The existence of a radiation impedance for the electromagnetic radiation from an oscillating charge is shown to imply a fluctuating electric field in the vacuum."[18] Candelas[19] shows that "...pressure fluctuations associated with these energy fluctuations would confer on the charge an irregular motion. This motion would represent a non-constant acceleration and so would also lead to a systematic reaction dampening force acting on the charge."[19] This supports Fulton's ad hoc suggestion that hyperbolic motion is unphysical.[13] Sciama concludes, "...the classical results regarding the radiation emitted by an electron and the radiative reaction force on an electron...can be understood in terms of the spectrum of the field fluctuations perceived by the charge."[12] Sciama suggests a borrowing mechanism where the radiated energy is borrowed from the vacuum field during periods of uniform acceleration and then given back during nonuniform acceleration.[12] Thus any consideration of the nature of a charged particle and of radiation production should include the charge's interaction with the zero-point energy. The resolution of the equivalence paradox may come with the development of quantum gravity theories[20] in which the zero-point energy plays a crucial role.

Any complete theory of electrodynamics must include the zero-point vacuum fluctuations and their interaction with matter. Boyer[1] shows that matter affects the zero-point fluctuations and they in turn feed back and affect matter. In fact, it appears that elementary particles can be viewed as organized coherences or spatial resonances in the zero-point sea.[21] Senitzky[22] shows that the charge's source field and the vacuum fluctuations are inseparably intertwined as "merely two sides of the same quantum mechanical coin."[22] This duality is likened to quantum mechanics' wave-particle duality. Sciama further supports this idea of charge vacuum synergy by noting that "it is not in general possible to divide the stress energy tensor into a 'real particle part' and a 'vacuum polarization part' in an unambiguous way."[12]

VACUUM POLARIZATION

Different types of particles interact differently with the vacuum zero-point energy.[23-25] In a first order model, protons, nuclei, and heavy ions generally produce a spherically distributed vacuum polarization with lines of polarization converging sharply to the particle (Figure 1). In fact, Greenberg[26] has demonstrated that if the nucleus becomes large enough, the intensity of the polarization precipitates real electron-positron pairs from the vacuum.

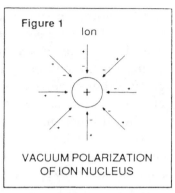

Figure 1

VACUUM POLARIZATION OF ION NUCLEUS

The spin of a particle also affects the vacuum fluctuations. Sciama notes that charged fields of differing spins (0, $\frac{1}{2}$, 1) give rise to different vacuum states.[12] Vorticity can also appear in the vacuum. Graham[27] has experimentally observed a macroscopic vacuum angular momentum caused by a static electromagnetic field's circulating Poynting vector. It is clear that different particles give rise to different vacuum interactions.

In view of this, can a conduction electron radiate differently than an ion? The quantum mechanical wave function description of the electron in matter or in a conductor is that of a smeared charge cloud. This smearing dilutes the vacuum polarization intensity and prevents the lines of polarization from converging onto the electron in a stable,

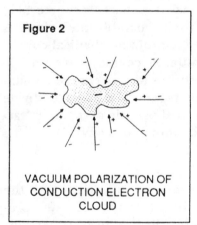

Figure 2

VACUUM POLARIZATION OF CONDUCTION ELECTRON CLOUD

orderly way (Figure 2). The electron could be described as a light, "ethereal" particle whose interactions with the zero-point fluctuations alter its form and actually cause it to smear. This intertwining interaction helps explain atomic ground state stability.[19] The electron tends to stabilize into "standing wave" harmonic eigenstates in matter. This smeared cloud is in equilibrium with the zero-point fluctuations. If we postulate there exist in the environment vacuum polarization displacement currents that can follow a particle's lines of polarization, then these currents would converge onto an ion but not converge onto a smeared electron cloud. Any environmental vacuum polarization displacement current would pass right through the smeared, fluctuating electron cloud. Senitzky shows that the "vacuum field plays no [net] role when the atomic system is an harmonic oscillator"[22] and that "linear oscillators such as antennas cannot in principle experience the effect of the vacuum field."[22] Also, Sciama shows that "for an [electron-based] detector at rest, the excitations caused by these zero-point fluctuations are precisely cancelled by its spontaneous emission rate."[12] Thus, the smeared electron cloud maintains thermodynamic equilibrium with the vacuum and could not absorb zero-point vacuum polarization surges.

However, the concentrated mass of the nucleus or heavy ion could interact with vacuum polarization modes, for its own vacuum polarized field convergently channels the longitudinal oscillations directly to the particle, *altering its momentum*. Note the two-way effect: the heavy particle can induce a spherically symmetric and convergent vacuum polarization around itself. As a transmitter, this field then launches vacuum polarization displacement currents whenever the particle moves or oscillates; as a receiver this field tends to channel those oscillations convergently onto the spherical particle, altering its motion. Because the heavy ion can maintain spherically convergent, stable lines of polarization, it becomes a transducer for transmitting and detecting longitudinal, vacuum polarization propagation modes that a conduction electron could not respond to and therefore could not detect.

Vacuum polarization effects can become powerfully synergistic when more than one ion or nucleus is involved. Roesel[28] describes the vacuum polarization potential for two extended charge distributions, and Soff[29] describes the

Moray B. King

"shake-off of the vacuum polarization cloud" in heavy ion collisions as a "collective type of electron-positron creation due to coherent action...."[29] Rauscher[30] was perhaps the first to suggest that coherent, vacuum quantum electrodynamic effects could take place in a plasma by demonstrating that vacuum polarization makes a significant contribution to the plasma's effective permittivity and conductivity. The nonlinear vacuum polarization description for a conglomerate of oscillating heavy ions would be quite complex and not readily solvable by the standard renormalization techniques. Modeling on a magnetohydrodynamic level would be more appropriate. In plasma analysis, the oscillations in polarization modulate the effective permittivity.[30] If similar modeling is applied to the vacuum's zero-point activity, the macroscopic, longitudinal, vacuum polarization oscillations could be described as "permittivity waves." A plasma model for the zero-point activity may yield a reasonable approximation, since Melrose[32] shows that the "vacuum polarization tensor in the presence of strong static homogeneous magnetic fields...reduces to forms equivalent to the magneto-acoustic and shear Alfven modes in a plasma."[32] Such modeling could predict longitudinal propagation modes. This could be reasonable, since Cover[31] demonstrates that vacuum polarization can give rise to longitudinal photon-like resonances. In this model, the highly nonlinear description of a group of ions interacting

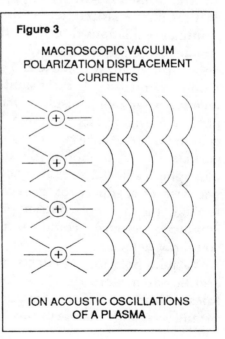

Figure 3

MACROSCOPIC VACUUM POLARIZATION DISPLACEMENT CURRENTS

ION ACOUSTIC OSCILLATIONS OF A PLASMA

with the vacuum energy could fulfill the nonequilibrium conditions identified by Nicholis, Prigogine,[33] and Haken[34] that give rise to self-organizing coherent behavior. If the ions in a plasma synchronously oscillate together, the intense vacuum polarization associated with the individual ions could coherently and synergistically add to give rise to a very intense macroscopic vacuum polarization (Figure 3).

ION-ACOUSTIC OSCILLATIONS

A natural place to look for evidence of macroscopic vacuum polarization would be in a plasma. The coherent oscillation of plasma ions is known as the ion-acoustic mode. Many investigators have observed that the ion-acoustic mode is associated with large radiant energy absorption,[35-37] vigorous high-frequency spikes,[38-42] runaway electrons,[43,44] rapid and anomalous plasma heating,[45-48] and anomalous plasma resistance.[48-51] Could any of the ion-acoustic anomalies be associated with the existence of macroscopic vacuum polarization effects?

There is evidence for ion-acoustic activity in nature's plasma. The evidence comes indirectly from the observation of whistler or sweeper waveforms. Whistlers[55] are waveforms that rapidly downshift their frequency and cannot normally be detected by standard narrow-band receivers. They are observed with increased ion-acoustic activity in laboratory plasmas.[38,51-54] Sweeping emissions are also observed in nature.[56-58] The following description of sweepers by Gerson[56] is similar to what Moray[60] described as the source of radiant energy driving his invention:

"Wideband noise bursts termed sweepers drift in frequency through portions of the HF and VHF bands. There are two broad types: (1) instantaneous, and (2) drifting mainly from higher to lower frequencies. They are readily observed at many locations over the planet. Their occurrence maximizes between 24–26 MHz. The instantaneous type is probably associated with thunderstorm activity. The drifting type may occur in trains that persist for hours. Individual members recur at closely the same time interval and display no significant dispersion. Intensities may be very high. They are generally not noticed when narrow band receivers are used. Their origin is not clear...." [56]

These sweepers may originate from tropospheric, ionospheric, magnetospheric, exospheric, and solar plasmas. [57,58] Podgornyi[59] notes that "the interplanetary medium is a giant reservoir filled with plasma in which various phenomena connected with collective interactions take place." [59] Webb[58] shows this activity persists through the atmosphere: "The geoenvironment is permeated with an electrical structure and with active electrical processes which serve to unify and control geoelectricity and to inter-relate geoelectricity with other physical aspects of the earth and its solar environment." [58] If the sweepers are associated with the vigorous ion-acoustic activity, then the action of the atmospheric, magnetospheric, exospheric, interplanetary and solar plasmas could be a source of longitudinal vacuum polarization displacement current. Could the ion-acoustic oscillations emit and detect this hypothesized form of radiant energy while conduction electrons not readily detect it?

MORAY'S DETECTOR

The hypothesis of the preceding section is viable, for the experiments conducted to confirm our present knowledge of electromagnetism has always used electron-based detectors. There is an exception, however. T. Henry Moray[60] experimented with ion oscillators and detectors, and as a result, he may have discovered what appears to be a novel energy source. Moray built a system of plasma tubes[61] and valves that were apparently tuned such that each tube resonated at its own ion-acoustic plasma frequency. The tubes were tuned and the circuit switching timed to shift the energy from the high-frequency stages down to the lower-frequency tubes. [62] A feature of resonating the ion-acoustic mode would be that the individual ions can experience a mutually coherent, low-loss harmonic motion without being totally disrupted by collisions. This would allow small pulses of energy from the previous stage to synchronously augment the oscillations. (Electrons are poor carriers for this purpose, since they are so light and their displacement is so large that they would undergo an excessive number of collisions per cycle of oscillation.) The ion-acoustic mode can also yield high electrical capacitance in each tube at its operating frequency. The oscillating ions in the plasma beget a maximal effective dielectric polarizability, while the anomalous resistance associated with the ion-acoustic mode prevents the plasma "dielectric" from breaking down. The use of many coupled stages at different operating frequencies allows a broadband interaction with the environmental energy. If impinging surges of vacuum polarization displacement currents encounter the ions in the tubes, the ions could synchronously move with them. Thus, Moray's ion-acoustic oscillators could resonate with the incoming vacuum polarization surges and absorb the energy.

Evidence to support the hypothesized existence of macroscopic vacuum polarization may come from studying the unusual characteristics of the output current from Moray's device. Most witnesses who observed the device in operation were impressed by the unusual, bright glow of the load-bank light bulbs. Another reported observation was that the conductive leads and the thin wires in the device remained cool even after hours of operation. This may be significant, for 30-gauge wire was used in the circuits within the device, and it delivered power on the order of kilowatts.[60] These observations might be explained by hypothesizing that the conductive leads acted as waveguides for the surrounding vacuum polarization displacement currents. In this case, the nuclei of the wire's metallic lattice would be the wave guiding structure with the conduction band electron cloud providing a smooth "continuity condition" to minimize scattering. Little net energy or momentum would be transferred to the conduction electrons since the vacuum polarization energy and the electron cloud are in thermodynamic equilibrium. The waveguide hypothesis can also be used to explain why a vigorous brush discharge was observed when the antenna was disconnected from the operating input-stage detector. Here, the detector itself set up high-frequency vacuum polarization displacement currents that were guided onto the antenna. This process would establish lines of polarization along the antenna that could then help channel environmental polarization displacement oscillations back to the detector's individual ions. This would augment the detector's effective cross-section for absorbing the environment's vacuum polarization energy. In this model, the ion oscillations and the vacuum polarization displacement currents are intimately phase-locked to yield a macroscopic wave-particle system. Perhaps Moray's invention was a manifestation of a macroscopic zero-point energy coherence.

The observations of the current from Moray's device suggest a qualitative experiment that could be used to help support the hypothesized significance of the ion-acoustic mode. A plasma tube is excited at its ion-acoustic frequency using an external power supply. If the hypothesized vacuum polarization displacement currents can be successfully coupled from the tube through conductive probes to an output load-bank circuit, then the current's characteristics can be compared to those that were observed in connection with Moray's device. If the output current during ion-acoustic resonance behaves similarly to the observed current from Moray's invention, and if the behavior cannot be duplicated by control tests using normal electrical conduction at the same power and frequency, then the tests would lend support to the hypothesis that the ion-acoustic mode launches macroscopic vacuum polarization displacement currents.

TESLA'S INVESTIGATIONS

The macroscopic vacuum polarization hypothesis may be applicable to explain Tesla's attempts to transmit and receive energy through high-potential devices (e.g., the tow-

Macroscopic Vacuum Polarization

ers at Colorado Springs[63] and Wardenclyffe[64]). The key transducer element in these structures would be the brush discharge corona around the sphere atop the towers. In order to allow coherent ion oscillations in this corona, it is most important to avoid sparking, for this will produce ion turbulence and disrupt the oscillations. Tesla avoided the sparking discharges by mounting smooth hemispherical capacitive structures at the top of his towers.[64] The circuit that energizes and couples to the corona would have to be tuned at the corona's ion-acoustic frequency. The tuning value is difficult to calculate, for the corona itself will increase the capacitance of the excitation circuit and alter its resonant frequency. Corona discharge studies[65-67] show that a stable, brush discharge corona can be induced by a unipolar radio frequency pulse burst. If the radio frequency matches the corona's ion-acoustic frequency, then a stable, coherent ion-acoustic oscillation can be maintained. This supports Corum's[68] suggestion that Tesla may have employed an X-ray ionization switch in order to achieve rectification of the driving radio frequency to induce and stabilize the corona atop his tower. If a stable ion-acoustic oscillation can be induced in the corona, then impinging vacuum polarization displacement currents could sympathetically couple to it and the energy could be absorbed into the coupled driving circuit.

Ion-acoustic oscillations may also be important for the production of ball lightning. Ball lightning[69,70] may be produced by replicating the bucking phase condition that Tesla associated with its production.[71] Ion-acoustic oscillations must first be induced in the corona around the Tesla coil. Then a signal or pulse must be abruptly switched into the circuit such that it is 180 degrees out of phase with the ion-acoustic oscillations. This bucking condition may induce a "vacuum polarization implosion" that could trigger the plasma to enfold into a vortex ring.[72-74] An anomalously long persistence of ball lightning triggered at relatively low energy levels could demonstrate the existence of macroscopic, coherent, self-organizing resonant states maintained by the zero-point energy.

SUMMARY

Many modern physicists have acknowledged that the zero-point energy or ether must be incorporated into any complete description of electromagnetic phenomena. Today's physics shows that different particles interact and polarize the vacuum in different ways. This suggests that ions can have different radiation characteristics than do conduction electrons. Even though today's physics cannot yet give solutions to the nonlinear multibody problem, it nonetheless recognizes the possibility that such nonlinear systems can manifest self-organizing coherent states. Since the individual ions in a plasma can coherently oscillate together, and since each ion exhibits an intense vacuum polarization, the ion-acoustic mode of a plasma may induce and detect macroscopic vacuum polarization displacement currents. The work of Moray and Tesla seems to support this hypothesis. It is hoped that this discussion will encourage experimental investigation of ion-acoustic oscillations in a brush discharge corona, for this nonlinear quivering transducer[75] may interact with coherent energetic modes in the zero-point sea.

ACKNOWLEDGEMENT

The author wishes to thank David Faust for helpful discussions.

REFERENCES

ZERO-POINT ENERGY ACTIVITY

1. T. H. Boyer, "Random Electrodynamics: The Theory of Classical Electrodynamics with Classical Electromagnetic Zero-Point Radiation," *Phys. Rev.* D 11 (4), 790 (1975).

2. C. Lanczos, "Matter Waves and Electricity," *Phys. Rev.* 61, 713 (1942).

Macroscopic Vacuum Polarization

3. E. G. Harris, *A Pedestrian Approach to Quantum Field Theory*. Wiley Interscience, NY, ch. 10, 1972.

4. C. W. Misner, K.S. Thorne and J.A. Wheeler, *Gravitation*, W.H. Freeman and Co., NY, ch 43-44, 1970.

5. A. Woodcock and M. Davis, *Catastrophe Theory*. Avon Books, NY, 1980.

UNIFORM ACCELERATION OF CHARGE

6. F. Rolrhich, "The Definition of Electromagnetic Radiation," *Il Nuovo Cimento* XXI, 811 (1961).

7. H. A. Atwater, "Radiation From a Uniformly Accelerated Charge," *Am. J. Phys.*, 38 (12), 1447 (1970).

8. V. L. Ginzburg, "Radiation and Radiation Friction Force in Uniformly Accelerated Motion of a Charge," *Sov. Phys. Uspekhi* 12 (4), 565 (1970).

9. D. G. Boulware, "Radiation From a Uniformly Accelerated Charge," *Ann. Phys.* 124, 169 (1980).

10. C. M. Dewitt and W. G. Wesley, "Quantum Falling Charges," *Gen. Rel. & Grav.* 2 (3), 235 (1971).

11. B. S. Dewitt and R. W. Breme, "Radiation Damping in a Gravitational Field," *Ann. Phys.* 9, 220 (1960).

12. D. W. Sciama and P. Candelas, "Quantum Field Theory, Horizons and Thermodynamics," *Adv. Phys.* 30 (3), 327 (1981).

13. T. Fulton and F. Rolrlich, "Classical Radiation From a Uniformly Accelerated Charge." *Ann. Phys.* 9, 499 (1960).

14. W. Pauli, Translated in *Theory of Relativity*, Pergamon Press, NY, p. 93, 1958.

15. R. Vasudevan, "Does a Uniformly Accelerated Charge Radiate?" *Lett. Al Nuovo Cimento* V (6), 225 (1971).

16. A. M. Luiz, "Does a Uniformly Accelerated Charge Radiate?" *Lett. Al Nuovo Cimento*, IV (7), 313 (1970).

17. C. Leibovitz and A. Peres, "Energy Balance of Uniformly Accelerated Charge," *Ann. Phys.* 25, 400 (1963).

18. H. B. Callen and T. A. Welton, "Irreversibility and Generalized Noise," *Phys. Rev.* 83 (1), 34 (1951).

19. P. Candelas and D.W. Sciama, "Is There a Quantum Equivalence Principle?" *Phys. Rev.* D 27 (8), 1715 (1983)

20. B. S. Dewitt, "Quantum Gravity," *Sci. Amr.*, 112 (Dec 1983).

VACUUM POLARIZATION

21. B. Toben, J. Sarfatti and F. Wolf, *Space-Time and Beyond*. E. P. Dutton and Co., NY. pp. 52-53, 1975.

22. I. R. Senitzky, "Radiation-Reaction and Vacuum Field Effects in Heisenberg-Picture Quantum Electrodynamics," *Phys. Rev. Lett.* 31 (15), 955 (1973).

23. F. Scheck, *Leptons, Hadrons and Nuclei*. North Holland Physics Publ., NY, pp. 212-223, 1983.

24. W. Greiner, "Dynamical Properties of Heavy-Ion Reactions - Overview of the Field," *S. Afr. J. Phys.* 1 (3-4), 75 (1978).

25. J. Reinhardt, B. Muller and W. Greiner, "Quantum Electrodynamics of Strong Fields in Heavy Ion Collisions," *Prog. Part. and Nucl. Phys.* 4, 503 (1980).

26. J. S. Greenberg and W. Greiner, "Search for the Sparking of the Vacuum," *Physics Today*, 24 (Aug 1982).

27. G. M. Graham and D.G. Lahoz, "Observation of Static Electromagnetic Angular Momentum in Vacuo," *Nature* 285, 154 (May 1980).

28. F. Roesel, D. Trautmann and R.D. Viollier, "Vacuum Polarization Potential for Two Extended Charge Distributions," *Nucl. Phys.* A 292 (3), 523 (1977).

29. G. Soff, J. Reinhardt, B. Muller and W. Greiner, "Shakeoff of the Vacuum Polarization in Quasimolecular Collisions of Very Heavy Ions," *Phys. Rev. Lett.* 38 (11), 592 (1972).

30. E. A. Rausher, "Electron Interactions and Quantum Plasma Physics," *J. Plasma Phys.* 2 (4), 517 (1968).

31. R. A. Cover and G. Kalman, "Longitudinal, Massive Photon in an External Magnetic Field," *Phys. Rev. Lett.* 33, 1113 (1974).

32. D. B. Melrose and R. J. Stoneham, "Vacuum Polarization and Photon Propagation in a Magnetic Field," *Il Nuovo Cimento* 32 A (4), 435 (1976).

SELF-ORGANIZING SYSTEMS

33. G. Nicolis and I. Prigogine, *Self-Organization in Nonequilibrium Systems*, Wiley, NY, 1977.

34. H. Haken, *Synergetics*, Springer-Verlag, NY, 1971.

ION-ACOUSTIC OSCILLATIONS

35. V. Yu. Bychenkov, A. M. Natonzon, and V. P. Silin, "Anomalous Absorption of Radiation on Ion-Acoustic Fluctuations," *Sov. J. Plasma Phys.* 9 (3), 293 (1983).

36. A. I. Anisimov, N. I. Vinogradov and B. P. Poloskin, "Anomalous Microwave Absorption at the Upper Hybrid Frequency," *Sov. Phys. Tech. Phys.* 18 (4), 459 (1973).

37. M. Waki, T. Yamanaka, H. B. Kang and C. Yamanaka, "Properties of Plasma Produced by High Power Laser," *Jap. J. Appl. Phys.* 11 (3), 420 (1972).

38. Yu. G. Kalinin, D. N. Lin, L. I. Rudakov, V. D. Ryutor and V. A. Skoryupin, "Observation of Plasma Noise During Turbulent Heating," *Sov. Phys. Dokl.* 14 (11), 1074 (1970).

39. H. Iguchi, "Initial State of Turbulent Heating of Plasmas," *J. Phys. Soc. Jpn.* 45 (4), 1364 (1978).

40. E. K. Zavoiskii, et al., "Advances in Research on Turbulent Heating of a Plasma," Proceedings of 4th Conference on Plasma Physics and Controlled Nuclear Fusion Research, pp. 3-24, 1971.

41. A. Hirose, "Fluctuation Measurements in a Toroidal Turbulent Heating Device," *Phys. Can.* 29 (24), 14 (1973).

42. V. Hart, private communication, 1982.

43. Y. Kiwamoto, H. Kuwahara and H. Tanaca, "Anomalous Resistivity of a Turbulent Plasma in a Strong Electric Field," *J. Plasma, Phys.* 21 (3), 475 (1979).

44. M. J. Houghton, "Electron Runaway in Turbulent Astrophysical Plasmas," *Planet. and Space Sci.* 23 (3), 409 (1975).

45. J. D. Sethian, D. A. Hammer and C. B. Whaston, "Anomalous Electron-Ion Energy Transfer in a Relativistic-Electron-Beam-Heated Plasma," *Phys. Rev. Lett.* 40 (7), 451 (1978).

46. S. Robertson, A. Fisher and C. W. Roberson, "Electron Beam Heating of a Mirror Confined Plasma," *Phys. Fluids*, 32 (2), 318 (1980).

47. M. Porkolab, V. Arunasalam, and B. Grek, "Parametric Instabilities and Anomalous Absorption and Heating in Magnetoplasmas," International Congress on Waves and Instabilities in Plasmas, Inst. Theoret. Physics, Innsbruck, Austria, 1973.

48. M. Tanaka and Y. Kawai, "Electron Heating by Ion Acoustic Turbulence in Plasmas," *J. Phys. Soc. Jpn.* 47 (1), 294 (1979).

49. Y. Kawai and M. Guyot, "Observation of Anomalous Resistivity Caused by Ion Acoustic Turbulence." *Phys. Rev. Lett.* 39 (18), 1141 (1977).

50. P. J. Baum and A. Bratenahl, "Spectrum of Turbulence at a Magnetic Neutral Point," *Phys Fluids* 17 (6), 1232 (1974).

51. M. Porkolab, "Parametric Instabilities and Anomalous Absorption and Heating of Plasmas," Symposium on Plasma Heating and Injection, Editrice Compositori, Bolona, Italy, pp. 46-53, 1972.

SWEEPING EMISSIONS, WHISTLERS

52. C. D. Reeve and R.W. Boswell, "Parametric Decay of Whistlers—A Possible Source of Precursors," *Geophys. Res. Lett.* 3 (7), 405 (1976).

53. M. S. Sodha, T. Singh, D. P. Singh and R. P. Sharma, "Excitation of an Ion-Acoustic Wave by Two Whistlers in a Collisionless Magnetoplasma," *J. Plasma Phys.* 25 (2), 255 (1981).

54. P. K. Shukla, "Emission of Low-Frequency Ion-Acoustic Perturbations in the Presence of Stationary Whistler Turbulence," *J. Geophys. Res.* 82 (7), 1285 (1977).

55. M. Watanabe, "On the Whistler Wave Solitons," *J. Phys. Soc. Jpn.* 45 (1), 260 (1978).

''' C. Gerson and W. H. Gossard, "Sweeping Emissions," *Phys.* :. 27 (4), 39 (1971).

57. S .R. P. Nayar and P. Revathy, "Anomalous Resistivity in the Geomagnetic Tail Region," *Planet. and Space Sci.*, 26 (11), 1033 (1978).

58. W. L. Webb, *Geoelectricity*, U. of Texas, El Paso, pp. 9-11, 1980.

59. I. M. Podgornyi and R. Z. Sagdeev, "Physics of Interplanetary Plasma and Laboratory Experiments," Sov. Phys. Uspekhi 98, 445 (1970).

INVENTIONS

60. T. H. Moray and J. E. Moray, *The Sea of Energy.* Cosray Research Institute, Salt Lake 1978.

61. T. H. Moray, "Electrotheurapeutic Apparatus," US Patent No. 2,460,707 (1949); contains corona discharge tubes.

62. M. B. King, "Stepping Down High Frequency Energy," Proceedings of the First International Symposium on Nonconventional Energy Technology, University of Toronto, pp. 145-158, 1981.

63. N. Tesla, *Colorado Springs Notes* 1899-1900, Nolit, Beograd, Yugoslavia, 1978.

64. N. Tesla, "Electrostatic Generators," *Sci. Amr.*, 132 (March 1934).

CORONA DISCHARGE, BALL LIGHTNING

65. W. W. Eidson, D. L. Faust, H .J. Kyler, J. O. Pehek and G. K. Poock, "Kirlian Photography: Myth, Fact and Applications," IEEE Electro 78 Convention Proceedings, Special Evening Session, Part One, SS/3, Boston, Mass., pp.1-21, May 23-25, 1978.

66. D. G. Boyers and W. A. Tiller, "Corona Discharge Photography," *J. Appl. Phys.* 44 (7), 3102 (1973).

67. D. Faust, private communication, 1984.

68. J. Corum, "Theoretical Explanation of the Colorado Springs Experiment," The Tesla Centennial Symposium, Colorado College, Colorado Springs (Aug 10-12, 1974).

69. J. D. Barry, *Ball Lightning and Bead Lightning*, Plenum Press, NY, 1980.

70. S. Singer, *The Nature of Ball Lightning*, Plenum Press, NY, 1971.

71. H. W. Secor, "The Tesla High Frequency Oscillator," *Electrical Experimenter* 3, 615 (1916).

72. W. H. Bostick, "Experimental Study of Plasmoids," *Phys. Rev.* 106 (3), 404 (1957).

73. D. R. Wells, "Dynamic Stability of Closed Plasma Configurations," *J. Plasma Phys.* 4 (4), 654 (1970).

74. P. O. Johnson, "Ball Lightning and Self-Containing Electromagnetic Fields," *Am. J. Phys.* 33, 119 (1965).

75. N. Tesla, *Lectures, Patents and Articles*, Nicola Tesla Museum, Beograd, Yugoslavia, pp. L65-68, 1956. (Reprinted by Health Research, Mokelumne Hill, CA 1973.)

6

The Adams Pulsed Electric Motor Generator

from

NEXUS Magazine

The Adams Pulsed Electric Motor Generator

a free energy machine at last?

Robert Adams
46 Landing Road
Whakatane
Bay of Plenty, New Zealand

THE REAL MCCOY

It is with great excitement, and appreciation to the inventor, that Nexus publish the following information on the Permanent Magnet Electric D.C. Motor Generator of Robert Adams, a former Chairman of the Institute of Electrical & Electronics Engineers, Inc., U.S.A., (N.Z. Section).

After having his invention suppressed for over 20 years, Mr Adams, at the age 72, has decided to share his design with the world regardless of the consequences.

Mr. Adams' quest to bring "free" energy to the world has cost him dearly, as it has many other researchers who threaten to bring the "establishment" undone.

He has survived an attempt on his life by an individual affiliated with the New Zealand Secret Intelligence Service and the Central Intelligence Agency, direct suppression of his invention by former (and recently deceased) Prime Minister of New Zealand, Robert Muldoon, the giant British electronics company, Lucas Industries, as well as numerous other insurmountable difficulties that have been placed in his path. All because his invention worked.

And not only that, it is so simple, any electronics manufacturer or skilled backyard-home-scientist could build one.

INVENTORS BEWARE !

In 1978 Mr. Adams discovered that inventors of machines or devices of high energy efficiency capability ("Free" Energy) are not only refused patents, but that in most cases, their inventions are classified under the "Military Use Clause", which is, of course, international. Inventors are prohibited from publishing details of their devices or promoting them in any manner if their invention is classified under this clause. In other words, their devices automatically become the sole property of the "establishment".

The fact that there is an established mechanism to suppress energy inventions of this nature has been a closely guarded secret for many years. Many inventors have made such claims, but the general public remain oblivious to the fact that they are being deprived of clean and free energy by organisations that would rather make money and hold power over the public, than allow such technology to become widely available. Yet another example of the abuse of power. (No pun intended.)

"FREE" ENERGY

This motor generator would be called a "Free Energy" machine by most individuals. It is, in fact, a device that converts the perpetual motion of sub-atomic particles, known in physics terminology as "particle spin", into conventional electric power. It is a widely accepted fact of physical law that sub-atomic particles are in a state of perpetual motion. Anybody who tells you that there is no such thing as perpetual motion is either ignorant or a liar. As Robert Adams states, "Our universe is a sea of energy - free, clean energy. It is all out there waiting for us to set sail upon it."

Adams has built a number of permanent magnet electric D.C. motor generators based on the principle outlined in this article, some of which have demonstrated an electrical efficiency of 690% and a mechanical efficiency of 620%. The devices run at room temperature. Any device that doesn't could not be running at over 100% efficiency, as heat is the major result of hysteresis losses that are induced in any conventional electric motor or generator. Radiated heat is a sure-fire sign that a power generator is not running over unity, as all heat radiated by such a device is wasted energy.

I will remind readers once again at this point that Mr Adams is not a fly-by-night, propeller-head, whacko, techno-boffin. He is an electrical engineer with over 60 experience in the field of electrical engineering, which has included designing and building equipment for use in power stations, broadcasting facilities, airport communications centres, etc. He is a former Chairman of the Institute of Electrical & Electronics Engineers, Inc., U.S.A., (N.Z. Section), and his resume includes personal referees ranging from a former New Zealand Commissioner of Police, a former Chairman of Air New Zealand, (and several technical specialists from the airline), as well as an Ex- NASA scientist.

Nexus would recommend to anybody interested in, or presently building a device of this nature, to try building a device based on Mr. Adams plans. His machines have demonstrated the ability to generate free energy, unlike most of the theoretical models that are promoted as over-unity devices.

GENERAL DESCRIPTION

The invention may be broadly said to be, an electric motor and/or generator comprising a rotor consisting of a number of radially arrayed permanently magnetised poles, and a stator consisting of a number of radially arrayed permanently magnetised poles, together with a number of wound poles.

The rotor's permanently magnetised poles use ferrite magnetic cores, and may comprise any even number of poles. The stator's wound poles employ steel or iron cores.

The device is essentially a D.C. machine, but may be fed A.C. input with the use of a solid state converter.

The Rotor uses a number of similar polarity permanent magnetic poles, i.e., all-South or all-North.

A further set of wound poles are radially arrayed in the stator, and are arranged in such a manner as to be fed energy, that is excited by back E.M.F. energy, from the poles of the rotor. Associated circuitry is provided to feed the energy back to the drive poles of the motor.

The resulting characteristics of this design is that once the rotor is moved from the position of equilibrium, each pole is attracted to, or repulsed by the stator poles, but at a precise geometrical point with respect to them, the input current to the drive coils ceases.

As a result, the collapsing field current is in the opposite direction to the applied force, thus reversing the magnetic polarity of the stator coils. This forces the rotor poles away (reaction), which is the instantaneous response of a system to an applied force, and is manifested as the exertion of a force equal in magnitude, but opposite in direction to the applied force.

Pulsing the D.C. input current, overcomes losses generated in conventional motors. According to classical electrical engineering theory, efficiency is greater the more nearly equal the Back E.M.F. (electromotive force) is to the applied voltage, i.e. the lower the input current. Figure 6 shows that there is minimum 100% back E.M.F. relative to the supply source of input D.C. voltage (according to classical electrical theory), which virtually depicts a sine wave due to the effect of the collapsing field.

This effect also overcomes the electrodynamic torque problems associated with conventional motor designs. (As input power varies with the duty cycle pulse; i.e. the lower the input current, the lower the input current, and the lower the speed, the greater the torque.) At clip-off, the back-EMF ceases, the collapsing field takes over, opposing the outgoing rotor magnet and thus increasing momentum.

With this design force is applied twice during each D.C. pulse, with pulse-on, and with pulse-off.

The timing of the pulses are determined by the dimensions of the motor itself, i.e. the speed of rotation of the motor's central axle, the position of the rotor magnets in relation to the stator windings, as well as the distance that the rotor magnets travel when passing across the poles of the stator winding (See accompanying diagrams).

CONSTRUCTION AND OPERATIONAL NOTES
Important Factors

1) Care must be exercised when assembling and wiring the drive windings to make sure that their polarities match the rotor magnet polarity.

2) Common earthing must be avoided in order to preclude

voltage and /or current loops. (If a number of drive windings do need to be commoned, use very low resistance conductors and employ a transmission type earthing system only.)

Stator (Drive) winding resistances are your choice. Robert Adams' machines were built varying from 0.03125 to 27 ohms per set. He has experimented with two, four, and eight pole machines. Efficiency increases with the number of wound poles in the stator.

Motor generators with a single, two, or three phase can be built to this design. A number of rotors may be ganged together on the same shaft in order to increase power output and does not require the use of any commutator, brushes or slip rings, all of which contribute to energy losses in ordinary motor generators.

Unlike conventional Series D.C. machines, this motor can be off-loaded, finds its own speed, and will run at that speed indefinitely. A conventional DC motor will run itself to destruction with off-loading. It requires no cooling, nor any overload protection, even if short circuited.

A number of highly qualified individuals have seen these devices running and producing energy at well above 100% efficiency. Let's hope that some of you can achieve similar results.

GENERAL CONSTRUCTION AND TESTING PROCEDURES OF THE ADAMS MOTOR GENERATOR

An ideal drive winding pole can be very readily available by obtaining some B.P.O. 3000 type relays (ex Telecom). Simply remove present winding, cut core in half, re-thread, assemble and fill with winding. This is a quick and cheap method of obtaining a very high quality non-retentive steel core. As aforementioned, winding resistances used by the inventor varied between 0.03125 to 27 ohms.

The above windings described are ample to drive prototypes even in a 180° application. You will find speeds up to 2500 rpm with only two of these windings 180° apart - no problem.

For A.C. Output Coil Windings and Core:

Ideal cores can be built cheaply and quickly by dismantling a spare power or audio transformer and utilising the "I" section laminations, obtain winding former to fit same and it is ready for winding. Turns and gauge will depend on what voltage and current you choose. Remember, at this stage, you should only be building a demonstration model, so to speak.

After a few changes, corrections and/or general modifications you will be ready to put a mechanical and/or electrical load on the machine. For an electrical load it is suggested you firstly wire up a bank of 6 - 12 LEDs. If everything is go, then switch over to torch lamps: Later on with a bigger machine - car lamps, or maybe household lamps and a mechanical load simultaneously.

For Efficiency Testing

Milliamp meters are useless for this machine - do not use for testing. Use only high quality digital true RMS meters, with input power, for high accuracy, use only a high quality electronic wattmeter: These instruments measure extremely accurately any wave-shape. A good twin-beam oscilloscope is a must: So too a high quality electronic temperature-measuring instrument with appropriate probe.

Drill dead centre of one or both cores, as per drawing.

Probe must be good fit. If, after one hour of running on load and temperature is around 40°, that will indicate things are most likely working correctly.

Don't forget your ambient in Australia will be considerably higher than ours. Read the inventor's rotary and solid state efficiency measurement data sheet.

Rotor-stator air gap is not critical, but the closer the better.

As stated elsewhere, the stator pole faces, if desired, may be reduced to 25% of the rotor pole face area, hence large drive windings and high drive current is not required.

With care to detail, correct mathematical calculations and high quality instrumentation correctly utilised, incredible results can be expected. Study data submitted shows those results have been attained on several machines. Temperature of conventional machines internally reach boiling point after fifteen minutes running. Check the Adams Motor Generator after running on full load for 48 hours, or after fifteen minutes if you prefer not to wait that long. You will be very pleasantly surprised - I refer to maximum loading conditions, not free running.

Mechanical Loading Test:

A high quality strain measuring instrument must be used in the universal "pony brake" method of mechanical load testing.

RPM tests must also be obtained with a high quality tachometer and/or oscilloscope reading and use the universal equation to calculate mechanical machine efficiency.

Note: Very Important Factor:

As you increase duty cycle, current input will increase and efficiency will decrease. From random test sheet results I have chosen it is very clear what to expect upon increasing duty cycle.

Note:

The Adams Motor Generator is so efficient, so simple and consequently it's construction is such that it surpasses overwhelmingly anything before it, thus lending itself admirably to mass production.

One of a number of unique features of the Adams Machine is the fact that the same rotor poles are utilised simultaneously for driving the machine and generating output energy.

Construction Equation - Adams Motor Generator - 20/12/76

It was found, after considerable development work, that maximum electro-magnetic effect produced in the stalloy stacked generating pole windings occurred when the dimension of the mating end of the stacks were four times greater in area than the rotor magnet's pole area. Hence the overall design of the machine incorporates this derived ratio of one to four. (The Adams Equation, as applies only to the Adams Machine).

Feedback - 20/12/1976

The feedback, produced by the output generating coil, produces a polarity reversal normally resulting in large eddy current losses in conventional machines, but, in the Adams Machine, it is harnessed to develop further additional torque to the magnetic rotor. The larger the output generating coils the greater the torque delivered to the rotor.

Power Factor - 1/7/1976

There is no power factor loss because the Adams Machine runs in a condition of resonance. Therefore, the Power Factor Loss is zero.

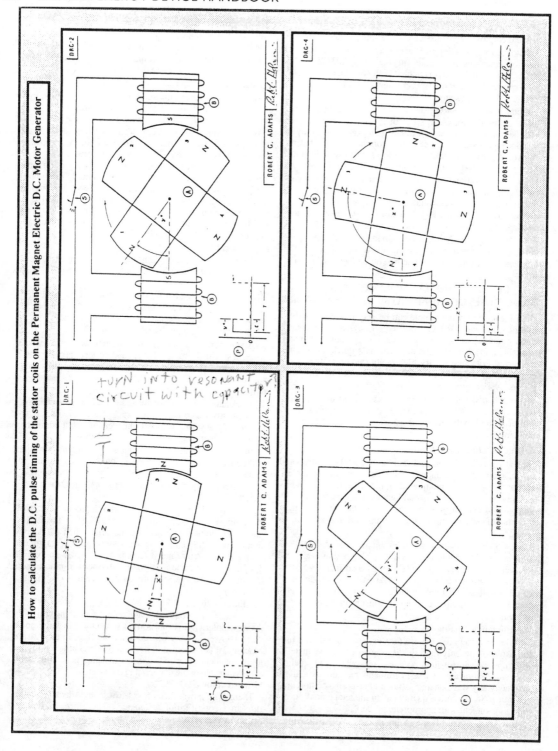

How to calculate the D.C. pulse timing of the stator coils on the Permanent Magnet Electric D.C. Motor Generator

tuYN iNto vesoNaNT circuit with capacitoY?

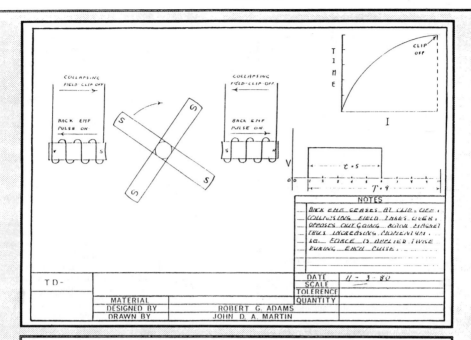

FIG. 5: Outline showing how double force pulse is produced by a single switched D.C. pulse.

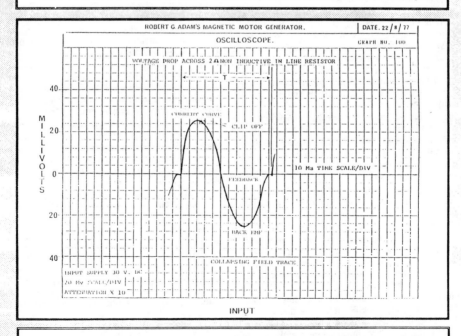

FIG. 6: Graph showing oscilloscope trace of characteristic voltage across stator windings.

I have for many years been waiting the opportunity to make a start on writing about my life as an electrical engineer and inventor, but have been reluctant to do so because of the possible reaction from one formidable individual who was instrumental in causing me untold frustration, ridicule, anxiety, financial problems, and health decline.

With the passing of Rob Muldoon, I now feel more deposed to exposing what he and his regime subjected me to, and the possible subsequent valuable loss to our country as a result.

I personally had a meeting with Muldoon and others at his home office in Tamaki regarding my invention.

The result of the meeting was that he recommended me to the Inventions Development Authority. In all good faith I duly contacted the inventions development authority and that folks was the beginning of the saga which was destined to follow.

The Inventions Development Authority passed me along to DSIR, who at the time were frantically working on their own energy systems, and assisting Government with various 'Think Big' projects.

Time rolled on, and DSIR monkeyed me around for several months; there were, as always, excuses for not proceeding with completion of their strange testing apparatus to test my machine.

In the meantime, Rob Muldoon appoints himself Minister in Charge of SIS, the department with world-wide connections to the CIA, ASIO, FBI and Interpol!

I had meanwhile, designed a bigger proving machine and had placed orders overseas for magnets and devices for pulsing equipment. Time went by, well beyond expected delivery dates, with no sign of the devices, nor any correspondence pertaining to my orders to any of the electrical companies. I lodged person-to-person calls to the people who originally signed the confirming correspondence to me, to be told that "that person is no longer with us and we cannot help you any more."

Muldoon had by now, committed the whole country to the New Plymouth Power Station, Huntly Power Station, Marsden Oil Refinery, etc etc. The Government signalled its intentions that it was in no way going to let a little outsider like me come into the act with a revolutionary machine capable of countering the so-called energy crisis.

Muldoon at this stage, had me well taped up so to speak. My phone, I was informed, was tapped, my mail was intercepted and I was kept under observation.

I held a number of meetings with the Chief Post Office Investigating Officer regarding the matter of mail disappearance and interception, with the same negative and unsatisfactory answers and results as from other Government departments. After the lapse of several months, the machine was finally evaluated by DSIR.

I must mention here an interesting fact regarding the evaluation of my machine by the DSIR. It happens that a certain electrical engineer that I was involved with, who incidentally, also evaluated my machine, informed me that the person designated the role of evaluating my machine in DSIR, was in fact a mechanical engineer with no background, knowledge, nor qualifications of any description pertaining to electrical or electronic engineering, and further that the

Auckland Division did not possess a member on the staff qualified to undertake such a project. This person none-the-less did do the evaluation and indeed signed the test results.

(You may well wonder how did this independent electrical engineer have such inside information on the DSIR? Well, it so happened that he himself had been their one and only electrical engineer in the Auckland Division, and I might add, a specialist on electrical motors too.)

That an unqualified person was assigned the task of evaluating the possibilities of a revolutionary motor is further evidence to me of conspiracy.

DSIR Evaluation - 5.8% efficiency

Lucas Industries Evaluation - 100% efficiency!

Other independent eminent engineers' findings, including those of the ex-electrical engineer from DSIR varied from 96.93% to 100% efficiency!

At this stage, I had unwittingly invited the then chief departmental district electrical engineer to be present at my laboratory to witness the phenomenon of one particular model displaying identical input and output wave forms on a twin beam oscilloscope with the machine windings running at ambient temperature, a condition which cannot be denied as proof of 100% efficiency, without any further tests being required.

He conceded there was no doubt whatever what he was witnessing was real, but like all academics who stick to their ivory tower scientific establishment beliefs, he said it could not be done.

Upon reading certain of my writings, it will be found that at one time in earlier years, I disbelieved in any kind of conspiracy regarding inventions pertaining to energy efficiency, and certain communication installations. I must now say, having unwittingly got myself into the web of the insidious conspiracy, through treading the path of an inventor in the field of free energy, that I now know from personal experience the pitfalls, stone walls, and blatant obstacles and barriers designed to hamper and silence inventors with such devices as energy efficient machines.

I decided to investigate the fate of a number of other excellent energy-efficient inventions, and learned that too many good inventions were never heard of again. Inventors themselves were turning into hermits, meeting with unexplained accidents, even totally disappearing. In many cases their laboratories were searched and ransacked, equipment confiscated and/or destroyed, and even attempts made on their lives. Others are frequently bought off in return for silence.

The conspirators, who are also the cartel operators are determined to continue to make mankind use fossil fuel for all possible energy requirements and will go to any lengths to achieve just that.

I have since superseded the above machine and have built and proven two different types of self-sustaining motor generators with efficiency ratings well beyond unity.

Use coil to produce An Opposing magnetic field in a permanent magnet, when the opposing field is maximum, switch off current, the collapsing magnetic field reenforced by the magnets field of the magn-

THE ADAMS PULSED ELECTRIC MOTOR GENERATOR
— UPDATE —

This invention by Robert Adams, a New Zealand inventor, is probably the nearest thing to a genuine free energy machine that NEXUS has encountered to date.

By Robert Adams

New Zealand
30th April 1993

In the April-May '93 issue of NEXUS, appended to the article, "Adams Breaks the Gravity Barrier" under the heading, 'Note for the Curious', I promised readers a set of drawings which would explain the questions regarding hysterisis, eddy current and magnetic drag losses as well as temperature ratings, etc.—these drawings to be accompanied with written explanations concerning the 'how' and 'why' of certain factors.

It is with my sincere regret that the above information and drawings will no longer be available to readers on an individual basis, but is briefly outlined in a modified version in this August-September issue of NEXUS.

Those readers who have written to me and sent monies over for copies of this information, are being returned their monies along with this standard announcement.

Subscriptions to this information are now closed as there is now sufficient information published in this article relating to this subject.

I wish my readers to know, however, that I will continue to publish certain other material in NEXUS relating to my inventions, but the certain 'how' and 'why' factors will be withheld due to circumstances beyond my control.

I am well aware, however, that this announcement will not prevent the intelligent and enthusiastic minds out there from continuing to probe and discover the remarkable secrets of the 'Adams PEMG' and continue to wish all my readers and correspondents the ultimate in success.

ROBERT ADAMS OF NEW ZEALAND OUTLINES MAGNETIC POLARITY REVERSAL AND HIS DISCOVERIES

As the inventor of the 'Adams Pulsed Electric Motor Generator', I write this treatise with a view to keeping it uncluttered from unnecessary theories and mathematics, so that all who read this article, whether they be enthusiasts, engineers or scientists, are able to follow the text, together with its drawings, describing the sequences in various stages of operation of the 'Adams Advanced PEMG'.

My various discoveries cover over twenty-five years in the fields of electrical rotary machines, with a total of over sixty years involvement in communications, broadcasting and electrical engineering.

It is my desire that as many free energy enthusiasts as possible get into the act of building my original machine whilst, at the same time, conducting their own research. Several people in different countries have already succeeded in building the machine in its original form, as has been outlined in the "Adams Manual", with beyond-unity results. After obtaining satisfying results from the original version, one would then be better equipped to handle the more stringent requirements of the 'Advanced', or 'Mark II' version. A lot of material in this treatise pertains to this 'Mark II' version, i.e., the 'Adams Advanced Motor Generator'.

LOSSES IN ELECTRICAL MACHINES

Losses in conventional electrical machines are too high and are due to magnetic drag, eddy currents and hysterisis, and consequent high operating temperatures.

It was with the above problems in mind that I was prompted to find a way of overcoming the aforementioned losses, the result of which evolved in a machine of beyond-unity capabilities.

As the 'Adams Motor' is a pulsed direct-current device, there is no change in polarity of the external source; therefore there are no eddy current losses, and hysterisis loss in the motor is minuscule; with new materials becoming available for stators, the small loss incurred would disappear. It matters little, however, as the machine efficiency is such that such a minuscule loss is negligible.

With reference to magnetic drag, this too is virtually non-existent, due to the unique design of the machine. A rotor pole, upon leaving the attraction area of the stator, is at a precise geometrical point—and is suddenly repulsed, thus overcoming any possibility of magnetic drag taking place.

Having outlined the above, I will now explain something regarding magnetic drag that won't be found in classical teachings or texts: a rotor, once in motion, is mutually attracted to any stator in its path. On leaving the stator area, however, the stator causes a drag-back effect—classical teachings do tell you that much. What classical teaching does not tell you is that the energy in the initial attraction equals exactly that which causes the drag upon leaving the stator area. This is where classical teaching is found wanting. The original attraction and secondary attraction exactly cancel one another out. Magnetic drag, therefore, does not exist in the straight motor version of the 'Adams Motor' invention. The machine is pulsed before the trailing edge of the rotor magnet can be affected. Should the timing be a little out, the effect would be minuscule.

Having discussed the aforementioned factors, there is little to explain regarding the very low operating temperature of the 'Adams Motor', as a result of which it does not require the use of a cooling fan with its own efficiency loss to counter.

It has been noted in *New Energy News* under the title "High Current Brushes", on making use of silver and platinum for commutator and points: this news is not news to me as, during my research in 1976, I used these materials myself for the above purpose. It appears, however, from this article submitted to *New Energy News* that the method for their use, being researched, has good prospects. I, for one, look forward to learning of the progress in this direction.

In 1976 I learnt of the high losses of my commutator system and first used silver for the star disc and platinum for the points with considerable success, and having since used photo and magnetic switching with remarkable success. Having, of course, kept my research and experimentation to myself over the past two decades for reasons associated in the main with the establishment, and latterly with international patent law, I was forced to withhold all my machine's secrets up until my recent decision to publish certain aspects of my work earlier this year.

RECENT DISCOVERY BY THE AUTHOR

One would expect magnetic polarity reversal to be instantaneous in a rotary machine between rotor magnet and stator. However, this is not so. I have recently discovered that the reversal is exponential in transition from one polarity to another. When this occurs, the magnetic radiation of the rotor pole/s doubles and, with no external power applied, there is no magnetism in the stator pole—so it becomes patently logical that the extra energy can only be from the ether (negative-time-energy).

ENGINEERING INTO NEGATIVE TIME AND NEGATIVE ENERGY

In the realms of engineering negative energy and negative time, I have anticipated there would be a new world of discoveries at hand and answers to be found to certain phenomena tak-

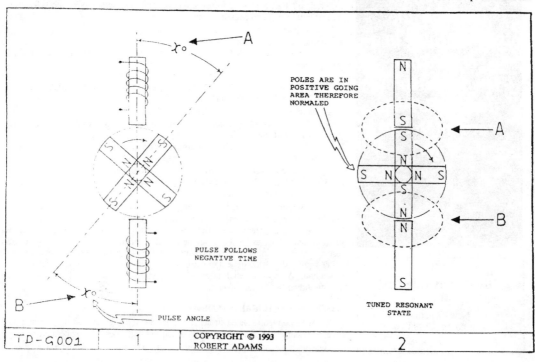

POLES ARE IN
POSITIVE GOING
AREA THEREFORE
NORMALED

PULSE FOLLOWS
NEGATIVE TIME

PULSE ANGLE

TUNED RESONANT
STATE

ing place, to which we have all previously been unaccustomed. This anticipation has manifested itself all too soon as, since my first successful recent attempts at engineering anti-gravity have proven, some interesting phenomena have become revealed, one of which is the process of magnetic polarity reversal, or conversion.

In an endeavour to discover what actually takes place during this 'conversion' of magnetic polarity, I used a magnetic polarity indicator and compass, but both proved worthless, as they simply hunted back and forth due to the pulses of magnetic fields from the machine being in motion.

Subsequent to this attempt, I had meanwhile made an important discovery concerning magnetic polarity reversal, in that it was not necessary that the machine be in motion or apply any external energy in order for it to bring about the magnetic polarity reversal. From this discovery, I became confident that I further discovered what actually takes place in this region in relation to the reversal phenomenon. In order to implement a test on this, I determined that a slow movement of the rotor by hand would indicate, on the instruments, what would be taking place. This exercise did indeed prove to be of substance, and I will now, here, graphically portray the results.

To deliver power, however, from the negative energy/negative time region, the machine must be in motion and, preferably, operating at certain harmonic speeds. The accompanying drawings give a physical description of the magnetic actions taking place as the rotor magnet reaches the stator and commences to traverse through the negative energy/negative time region.

In a 180° configuration, with two stator poles, the same actions take place simultaneously in reverse magnetic polarity order (as Drawing TD-G001, Figure 2, pointers A and B indicate).

ACTIONS TAKING PLACE IN THE 'ADAMS ADVANCED (MARK II) MOTOR GENERATOR'

1) A rotor south pole, upon approaching an open circuit stator, is mutually attracted to it, as depicted by Drawing TD-G004, Figure A.

2) When the leading edge of a south pole reaches the edge of a stator (Figure B), it appears, as it begins to move inwards, that the south polarity of the rotor pole is being exponentially reversed to north. In addition, the stator now becomes a temporary magnet, also exponentially becoming a north pole (Figure C). Meanwhile, the rotor magnet is still being attracted up to point zero of the stator (Figure D) and, as the leading edge of the rotor moves from point zero of the stator second-half region (Figure E), it appears that the north polarity strength is now increasing exponentially in this region and, on becoming parallel, i.e., reaching each other face-to-face (Figure F), the magnetic polarity reversal is then complete, and both magnet and stator poles are at <u>north</u> polarity. It is in this region now that the state of anti-gravity and negative time exists, with two magnets of like poles attracting each other and creating a gravitational repulsive force at the completion of the magnetic polarity reversal cycle.

There is a specific point of 'x°' from the stator centre where the machine is pulsed (refer Drawing TD-G001, Figure 1, A & B—pulse angle). Fine-tuning the timing at this geometrical point, the machine passes into a state of electromotive resonance where input power drops dramatically and shaft power increases in the negative time and negative energy region.

In all, the machine benefits from four different force actions per revolution and paying a minuscule toll fee for only one.

Firstly, as depicted by Figure A of Drawing TD-G004, the rotor magnet is mutually attracted to the stator (gets away with-

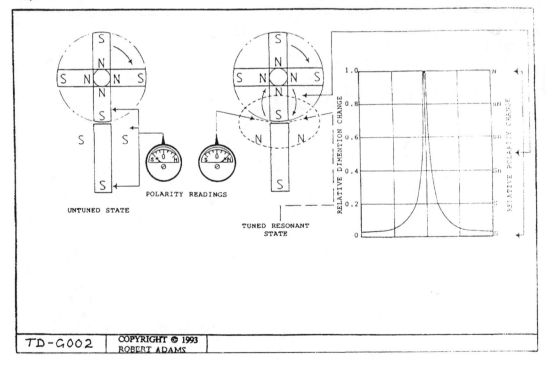

POLARITY READINGS

UNTUNED STATE

TUNED RESONANT STATE

out paying for that—explained elsewhere—refer paragraph 4 in section headed 'Losses in Electrical Machines'). Secondly, the attraction of the gravitational repulsion forces in the anti-gravity area (as illustrated in Drawing TD-G001, Figure 2, at arrows A & B). Thirdly, from the repulsion pulse of the stator at point 'x°' (refer to pulse angle of Drawing TD-G001, Figure 1). Fourthly, the rotor is given a further pulse from the collapsing field (a few degrees from point 'x°' in Figure 1).

For maximum possible results from the 'Adams Advanced (Mark II) Motor', it is necessary to apply harmonic/resonance equations for the calculation of all parameters including speeds and frequency. With the foregoing parameters met, it is recommended to engage magnetic or photo switching with its low loss, high efficiency properties. Drawing TD-G002, on the left, illustrates a positive 'untuned state' and, on the right of the same drawing, a 'tuned resonant state', together with a graph showing relative polarity change with component dimension changes. The area within the lower circle of the system indicates both poles are at north polarity (note the two curved arrows at each side of the magnet pole which depict the change that is/has taken place in the negative-time area).

It is possible to engineer the 'Adams Advanced (Mark II) PEMG' in such a way that a machine of any desired efficiency may be constructed from 100% up to four figures and beyond.

The term 'efficiency' now becomes a matter to be addressed, which I have done in the section headed 'Free-Energy Devices and the Term 'Efficiency' and its Connotations'.

PERMANENT MAGNETS AND 'WORK'

Permanent magnets do not and can not 'do work'—as claimed by certain people.

It is the ether/gravity forces which cause immense attraction and/or repulsion to take place between a permanent magnet and other magnetic material. It is these ether forces that, combining with the captive permanent magnet fields, harness the energy of gravity ether forces, so often erroneously referred to as 'work'—'done by magnets' (refer to notes on Nikola Tesla at the end of this section).

The magnets in this situation are simply acting 'as a gate', making way to the ether for the collection/release of gravitational/ether energy. The permanent magnet is a component in the system, operating as a 'gating device', as explained elsewhere in my writings—the magnet does not generate or create power (refer to Drawing TD-G005).

If magnets were doing 'work', they would heat up! The contrary takes place in negative-time systems during operation: rotor magnets drop in temperature below ambient in the above-described environment. Subtracting the drop in temperature of the rotor magnets from any small rise (if any) in stator temperature, due to minuscule hysterisis, would cancel the difference. The deeper the condition of resonance in the system, the lower the temperature of the magnets and stator windings.

When magnets and stators are engineered into negative time, the rotor, upon eclipsing face-to-face with the stator poles, causes a state of negative energy/negative time in that area at that moment in time. Almost simultaneously there is, in addition, the vector zero stress due to the resonantly-tuned wave trains of the stator pole generated voltage with that of the pulse voltage wave, resulting in a near mirror image.

The negative time/negative energy area between rotor magnet and stator independently causes an increase of 100% magnetic radiation every time a pole of the rotor passes a stator.

In this negative time, gravitational force is reversed; so in negative time, gravity becomes a repulsion force, not an attraction force.

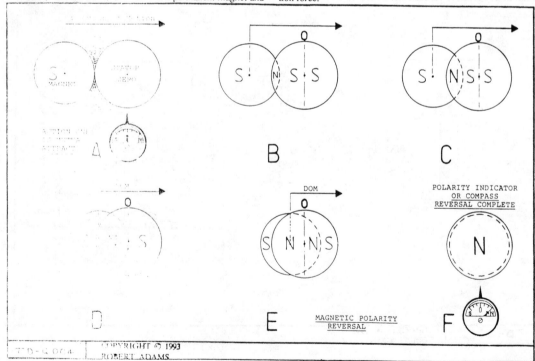

NEGATIVE TIME/NEGATIVE ENERGY RADIATION PATTERN

Irrespective of a magnet's gauss rating in a negative time/negative energy device, the area of magnetic radiation can be seen to double when a specific magnet and stator are engineered into 'negative time'. This radiation originates at the centre of the negative time region between the magnet and the stator, and spreads out radially and diminishes in strength as the square of the distance.

The enormous increase in radiation into space, and negative energy developed, is not generated by the magnet, as a lot of people would have you believe. Its source is (again) due to the magnet forming a gate and tapping gravitational energy with the result of gravitational repulsion, as explained in 'Engineering into Negative Time and Negative Energy'.

The area of radiation so covered by a negative time/negative energy device measures always an exact electromagnetic light gravity harmonic distance figure taken from point of origin.

These above statements are not theoretical but are indeed discoveries and results of actual tests and measurements undertaken in the laboratory by myself.

NIKOLA TESLA

While going through my notes and excerpts on Nikola Tesla's findings on Sunday 20 June 1993, I discovered a prepared statement of Tesla's 10 July 1937 work and another from the *New York Herald Tribune* dated 11 September 1932.

Tesla's statement below, dated 10 July 1937, vindicates completely my statement that magnets do not and can not 'do work'. Tesla's statement is:

"There is no energy in matter other than that received from the environment. It applies rigorously to molecules and atoms as well as the largest heavenly bodies and to all matter in the universe in any phase of its existence from its very formation to its ultimate disintegration."

"A few words will be sufficient in support of this contention. The kinetic and potential energy of a body is the result of motion and determined by the product of its mass and the square of velocity. Let the mass be reduced, the energy is diminished in the same proportion. If it be reduced to zero, the energy is likewise zero for any finite velocity. In other words, it is absolutely impossible to convert mass into energy. It would be different if there were forces in nature capable of imparting to a mass infinite velocity. Then the product of zero mass with the square of infinite velocity would represent infinite energy. But we know that there are no such forces and the idea that mass is convertible into energy is rank nonsense."

Nikola Tesla's statement of 11 September 1932, *New York Herald Tribune* is:

The assumption of the Maxwellian ether was thought necessary to explain the propagation of light by transverse vibrations, which can only occur in a solid. So fascinating was this theory that even at present it has many supporters, despite the manifest impossibility of a medium, perfectly mobile and tenuous to a degree inconceivable, and yet extremely rigid, like steel. As a result, some illusionary ideas have been formed and various phenomena erroneously interpreted. The so-called Hertz waves are still considered a reality, proving that light is electrical in its nature, and also that the ether is capable of transmitting transverse vibrations of

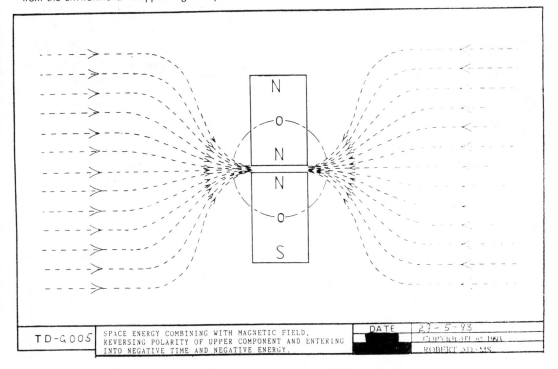

TD-G005	SPACE ENERGY COMBINING WITH MAGNETIC FIELD, REVERSING POLARITY OF UPPER COMPONENT AND ENTERING INTO NEGATIVE TIME AND NEGATIVE ENERGY.	DATE	23-5-93
			COPYRIGHT 1993
			ROBERT ADAMS

frequencies however low. This view has become untenable, since I showed that the universal medium is a gaseous body in which only longitudinal pulses can be propagated, involving alternating compressions and expansions similar to those produced by sound waves in the air. Thus, a wireless transmitter does not emit Hertz waves, which are a myth, but sound waves in the ether, behaving in every respect like those in the air, except that, owing to the great elastic force and extremely small density of the medium, their speed is that of light."

Although personal friends, Tesla and Einstein did not always agree with each other on certain points. However what Tesla is saying here is that though E=Mc², you can't simply choose a lump of mass such as a permanent magnet, place it in a system and extract energy from it. As aforesaid, in Tesla's own words, "the idea that mass is convertible into energy is rank nonsense."

FREE-ENERGY DEVICES AND THE TERM 'EFFICIENCY' AND ITS CONNOTATIONS

The term 'efficiency' or 'efficiency losses' relative to an external source, loses all substance of meaning when considering devices capable of well beyond unity. The term is no longer a yardstick, so to speak, as it becomes a relic of establishment teachings and present-day so-called conservation of energy laws, which now require to be rewritten. We must, therefore, now have a negentropy law.

Losses, if indeed any exist in a beyond-unity device, would be minuscule and of no substance, due to the output capacity of the device.

In my opinion, the most salient factor to look for in a device claimed to operate beyond unity, is its operating temperature under full load. This factor tells all, without the initial necessity to carry out exhaustive test procedures.

The matter of temperature of beyond-unity devices brings to mind Tesla's electric car. It is stated in my notes that the machine becomes very hot during operation. This, of course, is to be expected, as the 'free-energy section' of the machine is a

separate entity to the car motor proper, and in the year 1931, when Tesla tested his "Pierce Arrow" car, conventional DC motors were notoriously inefficient—around the order of 35%—and, incidentally, meanwhile, haven't improved that much. In addition, the confined space would have also been of no help, even with the assistance of a fan, which also had to be used according to his notes.

However, in contrast, my beyond-unity power device ('gravity generator') would be operating at least 20-40° Centigrade below ambient.

As the father of many discoveries and inventions pertaining to coils, transformers, pulsing systems and electric motors, on reflection it is unfortunate that Tesla hadn't figured out what could be done with his pulsing systems in relation to electric motors. Had he done so, he would not have required (according to Müller) powerful magnets or a cooling fan.

It is my opinion, after many years' experience in the free-energy research field, that a table of negative time/negative energy 'rating' be formulated in relation to devices using permanent magnets in free-energy applications. I am, at present, endeavouring to work out a system of magnetic radiation field strength measurement as a possible means of evaluating rotary devices that utilise permanent magnets. A system of this nature would distinguish 'beyond-unity shoptalk' from 'conventional shoptalk' and the term 'efficiency' would remain relegated to conventional devices. As the future 'beyond-unity empire' will grow and mature, so conventionalism, along with its present terminology, will wither and die.

The universe is negentropically organised and is proceeding transfinitely from disorder to order. This is not the concept of energy taught today in college and university campuses which persist in drumming in the long since foregone notions of Sir Isaac Newton and James Clerke Maxwell. We are, right this very moment, entering a new era of science, somewhat divorced from the trappings the establishment has been peddling for decades. We do not need to 'wait upon' the establishment to catch up to us here, for if we did we would still be a century behind in a century's time. It is for the establishment teachings

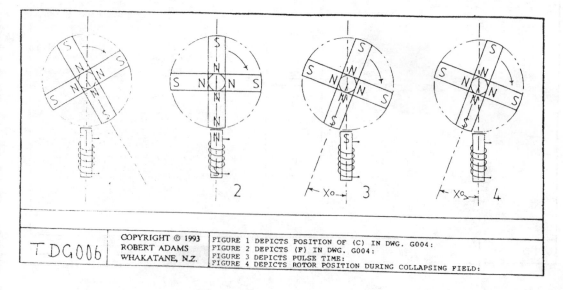

| TDG006 | COPYRIGHT © 1993 ROBERT ADAMS WHAKATANE, N.Z. | FIGURE 1 DEPICTS POSITION OF (C) IN DWG. G004: FIGURE 2 DEPICTS (F) IN DWG. G004: FIGURE 3 DEPICTS PULSE TIME: FIGURE 4 DEPICTS ROTOR POSITION DURING COLLAPSING FIELD: |

to latch onto us, which they will do—eventually—one day. In the meantime, new science will continue to forge ahead regardless, until sheer embarrassment will force classical teaching out the door.

With anti-gravity/beyond-unity devices, we must now go on to reconstitute our scientific laws and introduce a few new ones. As aforementioned, one being a negentropy law within which we must now go on to agree to the use of a more appropriate term of measurement for devices using permanent magnets in free-energy applications; and, as aforesaid, the term 'rating' comes to mind. For the purpose of this treatise I will now propose to use the term 'beyond-unity rating' or 'BUR' as an abbreviation, for the moment.

Beyond-unity devices' 'ratings' should, in my opinion, be according to an accepted table of values, ranging from what is, at present, termed 'unity' upwards, and thereby dumping the term 'efficiency' completely.

As there is no upper limit of negative energy other than 'blast-off' and/or 'self-annihilation' upon reaching absolute resonance, then a suitable table of values, with a suitable baseline, would be a practical solution. As the term 'unity' has been so indelibly engrained, then it would perhaps be suitable as a 'base'.

No doubt the establishment will 'perform' at my suggestions and/or recommendations on this—so be it. It would only take a small section of the researchers of new science amongst the 'beyond-unity circle' to agree upon the adoption of a new system of measurement and/or definition and publish it internationally along with the appropriate table of figure ratings. This would then enable beyond-unity researchers and adherents alike a more realistic platform as a base to work from, whilst still allowing the term 'efficiency' to apply to conventional below-unity apparatus.

This adoption of a new and separate system (or law) of the measurement of over-unity devices will, at the outset, identify and distinguish the subject of beyond-unity from its conventional counterparts and allow beyond-unity and anti-gravity researchers to get on with the job without harrassment from the classical thinkers.

7
Homopolar "Free-Energy" Generator

Homopolar "Free-Energy" Generator

Robert Kincheloe
Professor of Electrical Engineering (Emeritus)
Stanford University

Paper presented at the 1986 meeting
of the
Society for Scientific Exploration
San Francisco

June 21, 1986

Abstract

Known for over 150 years, the Faraday homopolar generator has been claimed to provide a basis for so-called "free-energy" generation, in that under certain conditions the extraction of electrical output energy is not reflected as a corresponding mechanical load to the driving source.

During 1985 the author was invited to test such a machine. While it did not perform as claimed, repeatable data showed anamalous results that did not seem to conform to traditional theory. In particular, under certain assumptions about internally generated output voltage the increase in input power when power was extracted from the generator over that measured due to frictional losses with the generator unexcited seemed to be about 26% of the maximum computed output power.

The paper briefly reviews the homopolar generator, describes the tests on this particular machine and summarizes the resulting data.

The Sunburst Homopolar Generator

In July, 1985, the author was invited to examine and test a so-called free-energy generator known as the Sunburst N Machine. This device was designed by Bruce DePalma and constructed with the support of the Sunburst Community in Santa Barbara, CA, about 1969. The term "free-energy" refers to the claim by DePalma[1] (and others[2]) that it was capable of producing electrical output power that was not reflected as a mechanical load to the driving mechanism but derived from presumed latent energy of a spatial magnetic field.

Apart from mechanical frictional and electrical losses inherent in the particular construction, the technique employed was claimed to provide a basis for constructing a generator which could supply the energy to provide not only its own motive power but also additional energy for external use. From August 1985 to April 1986 a series of measurements were made by the author to test these claims.

Generator Description

Details of the generator construction are shown in Figs. 1 and 2. It consists essentially of an electromagnet formed by a coil of 3605 urns of #10 copper wire around a soft iron core which can be rotated with the magnetic field parallel to and symmetrical around the axis of rotation. At each end of the magnet are conducting bronze cylindrical plates, on one of which are arranged one set of graphite brushes for extracting output current between the shaft and the outer circumference, and a second set of metering brushes

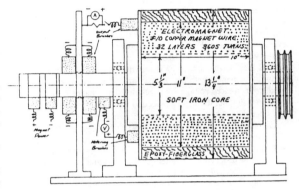

Figure 1 - Sunburst Homopolar Generator - side view

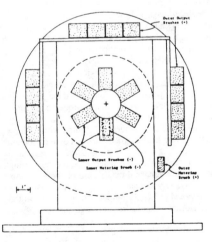

Figure 2 - Sunburst Homopolar Generator - Output (Brush) End View

The generator may be recognized as a so-called homopolar, or acyclic machine, a device first investigated and described by Michael Faraday[3] in 1831 and shown schematically in Fig. 3. It consists of a cylindrical conducting disk immersed in an axial magnetic field, and can be operated as a generator with sliding brushes extracting current resulting from the voltage induced between the inner and outer regions of the disk when the rotational energy is supplied by an external driving source. The magnitude of the incremental radial generated voltage is proportional to both the strength of the magnetic field and the tangential velocity, so that in a uniform magnetic field the total voltage is proportional to the product of speed times the difference between the squares of the inner and outer brush radii. The device may also be used as a motor when an external voltage produces a radial current between the sliding brushes.

There have been a number of commercial applications of homopolar motors and generators, particularly early in this century[4], and their operating principles are described in a number of texts.[5] The usual technique is to use a stationary magnet to produce the magnetic field in which the conducting disk (or cylinder) is rotated. Faraday found, however, that it does not matter whether the magnet itself is stationary or rotating with the disk as long as the conductor is moving in the field, but that rotating the magnet with the conducting disk stationary did not produce an induced voltage. He concluded that a magnetic field is a property of space itself, not attached to the magnet which serves to induce the field.[6]

DePalma claimed[7] that when the conducting disk is attached to a rotating magnet, the interaction of the primary magnetic field with that produced by the radial output current results in torque between the disk and the magnet structure which is not reflected back to the mechanical driving source. Lenz's law therefore does not apply, and the extraction of output energy does not require additional driving power. This is the claimed basis for extracting "free" energy. Discussions of the torque experienced by a rotating magnet are also discussed in the literature.[8]

for independently measuring the induced voltage between these locations. A third pair of brushes and slip rings supply the current for the electromagnet. A thick sheath of epoxy-impregnated fiberglass windings allow the magnet to be rotated at high speed.

Because the simple form shown in Fig. 3 has essentially one conducting path, such a homopolar device is characterized by low voltage and high current requiring a large magnetic field for useful operation. Various homopolar devices have been used for specialized applications[9] (such as generators for developing large currents for welding, ship degaussing, liquid metal magnetohydrodynamic pumps for nuclear reactor cooling, torquemotors for propulsion, etc.), some involving quite high power. These have been extensively discussed in the literature, dealing with such problems as developing the high magnetic fields required (sometimes using superconducting magnets in air to avoid iron saturation effects), the development of brushes that can handle the very high currents and have low voltage drop because of the low output voltage generated, and with counteracting armature reaction which otherwise would reduce the output voltage because of the magnetic field distortion resulting from the high currents.

From the standpoint of prior art, DePalma's design of the sunburst generator is inefficient and not suitable for power generation:

1. The magnetic field is concentrated near the axis where the tangential velocity is low, reducing the generated voltage.

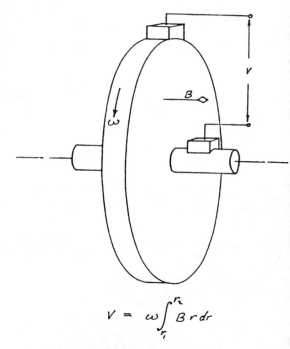

$$V = \omega \int_{r_i}^{r_i} B\, r\, dr$$

Figure 3 - Homopolar (Acyclic) Generator

2. Approximately 4 kilowatts is required to energize the magnet, developing enough heat so that the device can only be operated for limited periods of time.

3. The graphite brushes used have a voltage drop almost equal to the total induced voltage, so that almost all of the generated power is consumed in heating the brushes.

4. The large contacting area (over 30 square inches) of the brushes needed for the high output current creates considerable friction loss.

However, this machine was not intended as a practical generator but as a means for testing the free energy principle, so that from this point of view efficiency was not required.

DePalma's Results with the Sunburst Homopolar Generator

In 1980 DePalma conducted tests with the Sunburst generator, describing his measurement technique and results in an unpublished report[10]. The generator was driven by a 3 phase a-c 40 horsepower motor by a belt coupling sufficiently long that magnetic fields of the motor and generator would not interact. A table from this report giving his data and results is shown in Fig. 4. For a rotational speed of 6000 rpm an output power of 7560 watts was claimed to require an increase of 268 watts of drive power over that required to supply losses due to friction, windage, etc. as measured with the output switch open. If valid, this would mean that the output power was 28.2 times the incremental input power needed to produce it.

Several assumptions were made in this analysis:

PERFORMANCE OF THE SUNBURST HOMOPOLAR GENERATOR

machine speed:	6000 r.p.m.
drive motor current no load	15 amperes
drive motor current increase when N machine is loaded	1/2 ampere max.
Voltage output of N generator no load	1.5 volts d.c.
Voltage output of N generator loaded	1.05 v.d.c.
Current output of N generator (225 m.v. across shunt @ 50 m.v./1600 amp.)	7200 amperes
Power output of N machine	7560 watts = 10.03 H.p.

Incremental power ratio = $\frac{7560}{268}$ $\underline{28.2}$ watts out / watts in

Internal resistance of generator   62.5 micro-ohms

Reduction of the above data gives as the equivalent circuit for the machine:

R(int) 62.5 μΩ

N gen.

R(brush) 114.25 μΩ

R(shunt) 31.25 μΩ

1.5 v.d.c.

R(internal) = 62.5 micro-ohm
R(brush) = 114.25 " "
R(shunt) = 31.25 " "

BRUCE DEPALMA
17 DECEMBER 1980

Figure 4 - Test data from report by Bruce DePalma

1. The drive motor input power was assumed to be the product of the line voltage and current times the appropriate factor for a three-phase machine and an assumed constant 80% power factor. There was apparently no consideration of phase angle change as the motor load increased. This is clearly incorrect, since inclusion of phase angle is essential in calculating power in an a-c circuit, particularly with induction motors. It might also be noted that the measured incremental line current increase of 0.5 ampere (3.3%) was of limited accuracy as obtained with the analog clamp-on a-c ammeter that was used.

2. The output power of the generator was taken to be the product of the measured output current and the internally generated voltage in the disk less the voltage drop due only to internal disk resistance. Armature reaction was thus neglected or assumed not to be significant.

3. The generated voltage which produced the current in the main output brushes was assumed to be the same as that measured at the metering brushes, and the decrease in metered voltage from 1.5 to 1.05 volts when the output switch is closed is assumed to be due to the internal voltage drop resulting from the output current flowing through the internal disk resistance that is common to both sets of brushes and calculated to be 62.5 microohms.

Of these, the first assumption seems most serious, and it is the opinion of this author that some of DePalma's numerical results are questionable. A similar conclusion was reached by Tim Wilhelm of the Stelle Community in Illinois[11] who witnessed tests by DePalma in 1981.

Recent Tests of the Sunburst Generator by the Author

Being intrigued by DePalma's claims, the author accepted the offer by Mr. Norman Paulsen, founder of the Sunburst Community, to conduct tests on the generator which had not been used since the tests by DePalma.

Experimental Setup

A schematic diagram of the test arrangement is shown in Fig. 5. The generator is coupled by a belt to the drive motor behind it, together with the power supplies and metering both contained within and external to the Sunburst power and metering cabinet. The panel of the test cabinet provided power for the generator magnet and motor field. Meters on the panel were not functional and were not used; external meters were supplied. It was decided to use a d-c drive motor, primarily to facilitate load tests at different speeds and to simplify accurate motor input power measurements. The actual motor used was a surplus d-c generator from a DC-6 aircraft, rated at 400 amperes at 30 volts output from 3000 to 8000 rpm and capable of over 40 hp when used as a motor with appropriate forced air cooling. Half of the motor brushes were removed to reduce friction losses. Referring to Figure 9, variable d-c supplies for the motor armature and field and the homopolar generator magnet were provided by variacs and full-wave bridge rectifiers. Voltages and currents were mea-

sured with Micronta model 11-191 3½ digit meters calibrated to
better than 0.1% against a Hewlett Packard 740B Voltage Standard that was accurate to better than .005%. Standard meter shunts
together with the digital voltmeters were used to measure the
various currents. With this arrangement the generator speed could
be varied smoothly from 0 to 7000 rpm, with accurate measurement of motor input power, metered generator output voltage V_g
and generator output current I_g. Speed was measured with a
General Radio model 1531 Strobotac which was accurate to
better than 2%.

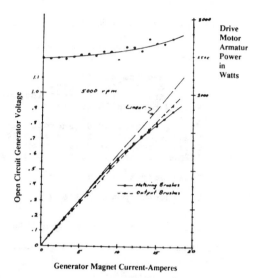

Figure 6 - Input power and generated voltage vs magnet current

Sunburst Homopolar Generator Test

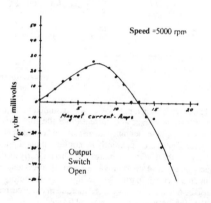

Figure 7 - Metering and output brush voltage difference vs magnet current

Figure 5 - Schematic diagram of generator test arrangement

Generator Tests

Various tests were conducted with the output switch open to
confirm that generated voltage at both the output brushes (V_{br})
and metering brushes were proportional to speed and magnetic
field, with the polarity reversing when magnetic field or direction
of rotation were reversed. Tracking of V_g and V_{br} with variation
of magnetic field is shown in Fig. 6, in which it is seen that the
output voltages are not quite linearly related to magnet current,
probably due to core saturation. The more rapid departure of V_g
from linearity may be due to the different brush locations as seen
on Fig. 3, differences in the magnetic field at the different brush
locations, or other causes not evident. An expanded plot of this
voltage difference is shown in Fig. 7, and is seen to considerably
exceed meter error tolerances.

Figure 6 also shows an approximate 300 watt increase in drive
motor armature power as the magnet field is increased from 0 to
19 amperes. (The scatter of input power measurements shown in
the upper curve of Fig. 6 resulted from the great sensitivity of the
motor armature current to small fluctuations in power line voltage,
since the large rotary inertia of the 400 pound generator does not
allow speed to rapidly follow line voltage changes). At first it was
thought that this power loss might be due to the fact that the outer
output brushes were arranged in a rectangular array as shown in
Fig. 1. Since they were connected in parallel but not equidistant
from the axis the different generated voltages would presumably
result in circulating currents and additional power dissipation.
Measurement of the generated voltage as a function of radial
distance from the axis as shown in Fig. 8, however, showed that
almost all of the voltage differential occurred between 5 and 12
cm, presumably because this was the region of greatest magnetic
field due to the centralized iron core. The voltage in the region of
the outer brushes was almost constant, with a measured variation
of only 3.7% between the extremes, so that this did not seem to

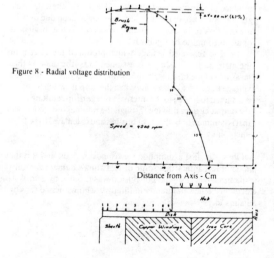

Figure 8 - Radial voltage distribution

Continued

explain the increase in input power. The other likely explanation seems to be that there are internal losses in the core and other parts of the metal structure due to eddy currents, since these are also moving conductors in the field. In any event, the increase in drive power was only about 10% for the maximum magnet current of 19 amperes.

Figure 9 typifies a number of measurements of input power and generator performance as a function of speed and various generator conditions. The upper curve (a) shows the motor armature input power for a constant motor field current of 6 amperes as the speed is varied with no generator magnet excitation and is seen to reach a maximum of 4782 watts as the speed is increased to 6500 rpm. This presumably represents the power required to overcome friction and windage losses in the motor, generator, and drive belt, and could be expected to remain essentially constant whether the generator is producing power or not.

Curve 14b shows the increase of motor armature power that results from energizing the generator magnet with a current of 16 amperes but with the generator output switch open so that there is no output current and hence no output power dissipation. This component of power (which is related to the increase of drive motor power with increased magnet current as shown in Fig. 6 as discussed above) might also be present whether or not the generator is producing output current and power, although this is not so evident since the output current may affect the magnetic field distribution.

Curve 14c shows the further increase of motor armature input power over that of curves 14a and 14b that results when the output switch is closed, the generator magnet is energized and output current is produced. It is certainly not zero or negligible as predicted by DePalma, but rises to a maximum of 802 watts at 6500 rpm. The total motor armature input power under these conditions is thus the sum of (a), (b), and (c) and reaches a maximum of 6028 watts at 6500 rpm.

The big question has to do with the generated output power. The measured output current 6500 rpm was 4776 amperes; the voltage at the metering brushes was 1.07 volts. Using a correction factor derived from Fig. 7 and assuming a common internal voltage drop due to a calculated disk resistance of 38 microohms, a computed internal generated potential of 1.28 volts is obtained which if multiplied by the measured output current results in an output power of 6113 watts. All of this power is dissipated in the internal and external circuit resistances, the brush loss due both to the brush resistance and the voltage drops at the contact surfaces between the brushes and the disk (essentially an arc discharge), and the power dissipated in the 31.25 microohm meter shunt. It still represents power generated by the machine, however, and certainly exceeds the 802 watts of increased motor drive power by a factor of 7.6 to 1. It even exceeds the input motor armature power of 6028 watts, although the total system efficiency is still less than 100% because of the generator magnet power of approximately 2300 watts and motor field power of about 144 watts which must be added to the motor armature power to obtain total system input power. It would thus seem that if the above assumptions are valid, DePalma correctly predicted much of the output power with this kind of machine is not reflected back to the motive source. Figure 10 summarizes the data discussed above.

	I	II	III	
MAGNET POWER OUTPUT SWITCH	OFF OPEN	ON OPEN	ON CLOSED	
SPEED	6500	6500	6500	RPM
MAGNET CURRENT	0	16	16	AMPERES
MOTOR ARMATURE POWER	4782	5226	6028	WATTS
INCREMENT		444	802		WATTS
METER BRUSH VOLTAGE	.005	1.231	1.070	VOLTS
OUTPUT CURRENT	0	0	4776	AMPERES
GENERATED VOLTAGE		1.280	(1.280)	VOLTS
GENERATED POWER	0	0	(6113)	WATTS

Figure 10 - Summary of test results at 6500 rpm

Homopolar Generator Test - Big Springs Ranch April 26, 1986

To further examine the question of the equivalence between the internally generated voltage at the main output brushes and that measured at the metering brushes, a test was made of the metered voltage as a function of speed with the generator magnet energized with a current of 20 amperes both with the output switch open and closed. The resulting data is shown in Fig. 11. The voltage rises to about 1.32 volts at 6000 rpm with the switch open (which is close to that obtained by DePalma) and drops 0.14 volts when the switch is closed and the measured output current is 3755 amperes, corresponding to an effective internal resistance of 37 microohms. Even if this were due to other causes, such as armature reaction, it does not seem likely that there would be a large potential drop between the output and metering brushes because of the small distance, low magnetic field (and radial differential voltage), and large mass of conducting disk material. Internal currents many times the measured output current of almost 4000 amperes would be required for the voltage difference between the outer metering and output brushes to be significant and invalidate the conclusions reached above.

A further method of testing the validity of the assumed generated output potential involved an examination of the voltage drop across the graphite brushes themselves. Many texts on electrical machinery discuss the brush drop in machines with commutators or slip rings. All of those examined agree that graphite brushes

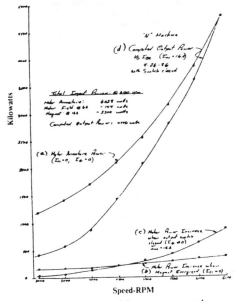

Speed-RPM

Figure 9 - Input and Output power vs speed

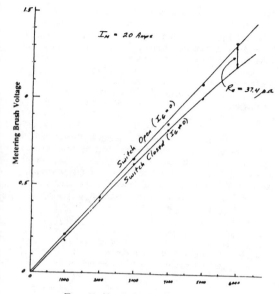

Figure 11 - Metering brush voltage vs speed

typically have a voltage drop that is essentially constant at approximately one volt per brush contact when the current density rises above 10-15 amperes per square centimeter. To compare this with the Sunburst machine the total brush voltage was calculated by subtracting the IR drop due to the output current in the known (meter shunt) and calculated (disk, shaft, and brush lead) resistances from the assumed internally generated output voltage. The result in Fig. 12 shows that the brush drop obtained in this way is even less than that usually assumed, as typified by the superimposed curve taken from one text. It thus seems probable that the generated voltage is not significantly less than that obtained from the metering brushes, and hence the appropriateness of the computed output power is supported.

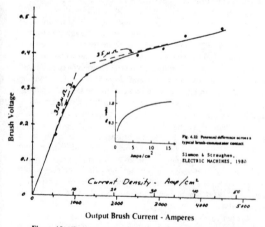

Figure 12 - Calculated output brush voltage drop vs current

Conclusions

We are therefore faced with the apparent result that the output power obtained when the generator magnet is energized greatly exceeds the increase in drive motor power over that required to supply friction losses with the magnet not energized, which is certainly anomalous in terms of conventional theory. Several possible explanations for this occur to the author:

1. There could be a large error in the measurements, such as some factor such as noise which caused the digital meters to read incorrectly or grossly incorrect current shunt resistances, although in the opinion of the author this seems unlikely.

2. There could be a large difference between the measured voltage at the metering brushes and the actual generated voltage in the output brush circuit due to armature reaction, differences in the external metering and output circuit geometry, or other unexplained cause, although again as discussed above the various data suggest that this is not likely.

3. DePalma may have been right in that there is indeed a situation here whereby energy is being obtained from a previously unknown and unexplained source. This is a conclusion that most scientists and engineers would reject out of hand as being a violation of known laws of physics, and if true would have incredible implications.

Perhaps other possibilities will occur to the reader.

The data obtained so far seems to have shown that while DePalma's measurement technique was flawed and his numbers overly optimistic, his basic premise has not been disproved. While the Sunburst generator does not produce useful output power because of the internal losses inherent in the design, a number of techniques could be used to reduce the friction losses, increase the total generated voltage and the fraction of generated power that can be delivered to an external load. DePalma's claim of free energy generation could perhaps then be examined.

It should be mentioned, however that the obvious application of using the output of a "free-energy" generator to provide its own motive power, and thus truly produce a source of free energy, has occurred to a number of people and several such machines have been built. At least one of these known to the author[12], using some excellent design techniques, was unsuccessful.

1. DePalma, 1979 a,b,c, 1981, 1983, 1984, etc.
2. For example, satellite News, 1981, Marinov, 1984, etc.
3. Martin, 1932, vol 1, p. 381.
4. Das Gupta, 1961, 1962; Lamme, 1912, etc.
5. See, for example, Bumby, 1983; Bewley, 1952; Kosow, 1964; Nasar, 1970.
6. There has been much discussion on this point in the literature, and about interpretation of flux lines. Bewley, 1949; Cohn, 1949a,b; Crooks, 1978; Cullwick, 1957; Savage, 1949.
7. DePalma, op. cit.
8. Kimball, 1926; Zeleny, 1924.
9. Bumby, Das Gupta, op. cit.
10. DePalma, 1980
11. Wilhelm, 1980, 1981 and personal communication.
12. Wilhelm, 1981

VI. The "N" MACHINES

d) Trombly-/Kahn-"N" Machine

The Trombly/-Kahn "N" Machine represents a new and improved version of classic "N" machine types. It is a departure from the past machines since it utilizes rotating electromagnets along with the central-"disc" rotating component.

The improved operation of the Trombly/-Kahn "N" Machine is achieved by providing a low reluctance magnetic return path for the magnetic flux that passes through the central rotor-(disc) component. This low reluctance return path permits the electromagnets to produce a high electrical field with a relatively small input current. Since the input current is low, overheating is avoided and the full potential of the homopolar generator is achieved.

The low reluctance magnetic return path is preferably produced by providing a relatively high permeability co-rotating enclosure (having enclosure halves) of sufficient radial and axial dimensions to enclose the electromagnets and disc conductor of the rotor. The disc conductor is preferably constructed from a high permeability, low resistivity material such as iron, and can be integral with the electromagnetic cores.

An International Patent has been issued on this present art which is listed as:-X02K31/00, and fully describes the details and specifics of the new type of "N" machine.

In the summary of this Patent, it is stated that this present invention provides a co-rotating homopolar generator that avoids the heating problems of prior machines and renders possible and convenient the generation of electricity at extremely high efficiency. The generator has a rotor comprising a disk conductor and co-rotating co-axial electromagnets on either side. The present invention achieves the improved operation by providing a low reluctance magnetic return path for the magnetic flux that passes through the disk conductor.

The low reluctance path permits the electromagnets to produce a high field-(limited to 2.2 Teslas by the saturation of iron) with a relatively low value of coil excitation current. Thus overheating is avoided and the full potential of the homopolar generator is achieved.

In the preferred embodiment, the low reluctance magnetic return path is provided by a relatively high permeability co-rotating enclosure,-(designated a "flux return enclosure") of sufficient radial and axial size to enclose the magnets and disk conductor of the rotor. Additionally, the disk conductor itself is preferably constructed from a high permeability low resistivity material such as silicon iron, and can indeed be integral with the electromagnet cores.

Output power is drawn between the periphery of the disk conductor (within the flux return enclosure) and the rotor shaft through fixed disk and shaft brushes. The disk brush protrudes through an annular slot in the flux return enclosure, and is geometrically configured so as not to add a large amount of reluctance to the flux return path. To this end, the disk brush is formed with a relatively thin web portion that passes through the enclosure gap. The web portion still has sufficient thickness so that the mechanical strength of the brush is not compromised. Moreover, the web portion has sufficient thickness, and hence conductance, that the saving in magnet power is not offset by excessive ohmic heating in the web portion.

A quote from Bruce DePalma about this latest Trombly/-Kahn "N" Machine,:-"Trombly and Kahn are two of the brightest young physicists in America today." "Their work is of the very highest quality and is described in the subject Patent and report."

Bruce DePalma believes that this new "N" Machine can be enlarged to handle more current by the application of a liquid metal brush system, such as he employs in his "N" Machine project work.

e) Bruce DePalma DePalma Institute, Santa Barbara, California

Although the concept of the "N" machine is not new, having been based on Faraday's disc of 1831, various researchers, including Bruce DePalma have made continuous performance improvements on these machines.

It was Michael Faraday who first rotated a copper disc between poles of a horseshoe permanent magnet and discovered that a voltage is produced between the central shaft and the outer edge of the disc. The disc has become known as the Faraday Homopolar-Disc generator, and the EMF is drawn off by brushes in contact with the shaft and outer disc edge.

The "N" machine basiclly consists of a high speed cylindrical permanent magnet from which electrical current (positive charge) with the circuit connection with brushes made in the same manner as the Homopolar Disc generators. These electrical generating units have been the first and simplest machines which exhibit an over-unity output, but only at very high speed levels of approximately 7000 rpm, and higher.

The "N" machines produce a uniform wattage flow at low voltage high current, which is a useful feature for many power applications. The requirement for the peripheral brush contacts has been, and to some extent remains a problem for these units due to the extremely high surface contact velocities.

Bruce DePalma has been active in evolving solutions to the various problems of the "N" machines, including the safeguarding against bursting of the rotating magnet at high speeds. One of his machines is essentailly a hybrid design,

CLOSED PATH - HOMOPOLAR GENERATOR
TROMBLY-KAHN-"N" MACHINE

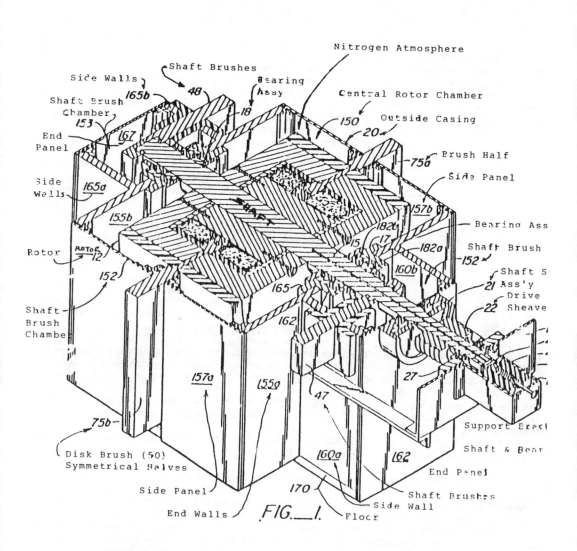

Nitrogen Atmosphere

Shaft Brushes

Side Walls

Shaft Brush
Chamber
153

End
Panel

Side
Walls

Rotor

Shaft
Brush
Chamber

Bearing
Assy

Central Rotor Chamber

Outside Casing

Brush Half

Side Panel

Bearing Ass

Shaft Brush

Shaft S
Ass'y
Drive
Sheave

Support Brac

Shaft & Bear

End Panel

Shaft Brushes

Side Wall

Floor

Disk Brush (50)
Symmetrical Halves

Side Panel

End Walls

FIG. I.

Nomenclature

12-Rotor
15-Shaft
17&18-Bearing Assemblies
20-Casing
21-Shaft Seal Assembly
27-Shaft End Bearing
30-Central Disk Conductor
32, A, B-Electromagnet Coils

35,A,B-Iron Cores
37,A,B-Flux Return Enclosure
40-Electrically Isolated Shaft Portion
42-Magnet Excitation Power Supply
45-Magnet Brushes
47-Shaft Brushes
48-Shaft Brushes
49-Disk Brush

VI. *The "N" Machines*

which features a Faraday copper disc combined with central ring magnets as the negative pole component. This machine ran at 7000 rpm, and above, producing an over-unity output.

One of the earlier problems with "N" machines has been the low-friction transfer of the high current from the high speed rotor through brushes of some special type. Liquid metals, such as Mercury have been used as the stationary current transfer means, but since Mercury is both costly and toxic in use, some "N" machine researchers prefer not to use it.

f) *Tom Valone* Integrity Electronics & Research, Buffalo, N.Y., 14221

Tom Valone has been actively involved in the continuing research and development of "N" Machines for many years, and has contributed to their improvement and acceptance. Like Bruce DePalma, Tom Valone believes in the value and advantages of the liquid brush system, and is currently using liquid solder for this prototype units.

From Tom Valone, June, 1985:-"In the field of non-conventional energy generators, the one-piece Homopolar of Faraday Generator has kindled a lot of attention. Ever since 1831, the rotating magnet and disc combination has defied complete analysis because of its operation totally within a non-inertial reference frame.

Conventional physics has attempted to come to terms with its operation, but anomalies still seem to remain.

A few physicists have recognized that its basis is relativistic, as can be seen by analyzing the polarization (electrical) set up by a moving body in a magnetic field. (Special Relativity). However without restrictions on mass and rotational speed, the only valid physical treatment seems to lie in the generally-covariant form of Maxwell's equations which can be applied in a non-inertial reference frame, (General Relativity). Einstein comes the closest to answering the questions about Faraday's Generator, by some still remain..."   T.V., 6/85.

Tom Valone's book, "The One-Piece Faraday Generator: Theory and Experiment" is available from:- Integrity Electronics & Research,-558 Breckenridge Street, Buffalo, New York 14222. 17.00 p.p.

VII. *MOTOR/GENERATOR UNITS & SYSTEMS ,*
  a) *Raymond Kromrey - Switzerland (1968)*

The electric generator (U.S. Pat. No. 3,374,376) has been designed to negate the effects of back-EMF within the field windings of generators by the application of special arrangement of field permanent magnets and an armature consisting of two series-connected coils.

The unit operates as a conventional, but opposed, two pole generator since the permanent magnetspoles are reversed at either end of the stator. The armature is thus demagnetized and remagnetized successively as it is revolved within framework bearings, with resulting reversal of polarity-and an A.C. output.

When the output circuit is open, the mechanical energy applied to the rotor/armature is converted into the work of magnetization, and when the circuit is closed, part of this work is converted into electrical energy as the current flowing through the windings opposes the magnetizing action of the field and increases the magnetic reluctance of the armature.

This above action, when joined with the flywheel effect provided by a coupled flywheel explains why the speed of this generator remains substantially unchanged when the output circuit is either opened or closed. As the armature approaches its position of alignment with the gap, the constant magnetic field exisitng there-across tends to accelerate the rotation of the armature relative to the pole pieces, thereby aiding the applied driving torque; the opposite action, ie: a retarding effect occurs after the armature passes through its aligned position. As the rotor attains a certain speed, however, the flywheel effect of its mass overcomes these fluctuations in the total applied torque so that a smooth rotation ensues.

The magnetic flux path includes two axially spaced magnetic fields traversing the rotor axis sustantially at right angles, these fields being generated by respective pole pairs co-operating with two axially spaced armatures of the character described. It will be generally convenient to arrange the two armatures in a common axial plane, the two field producing pole pairs being similarly co-planar.

The armatures are preferably of the laminated type to minimize the flow of eddy currents therein, thus, they may consist, in essence, of highly permeable (eg: soft iron)) sheets whose principal dimension is perpendicular to the rotor axis. If the ferromagnetic elements are part of the rotor, the output circuit will include the usual current-collecting means, such as slip rings or commutator segments, according to whether alternating or direct current is desired. The source of coercive force in the stator includes, advantageously, a pair of oppositely disposed yoke-shaped magnets, of the permanent or the electrically energized type, whose extremities constitute the aforementioned pole pieces. If electromagnets are used in the magnetic circuit, they may be energerized by an external source or by direct current from the output circuit of the generator itself.

In summation, the converter consists of an input drive motor which is directly coupled to this special type of generator which continues to run under load when the generator is short circuited. In general, the converter can be described as a single-phase motor-generator with a powerful permanent magnet stator and a rotor core of soft iron.

The "N" Effect

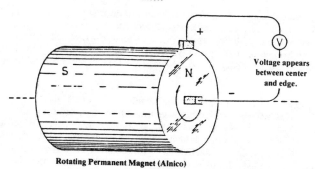

Voltage appears
between center
and edge.

Rotating Permanent Magnet (Alnico)

FIG. 22-B
THE "N" MACHINE

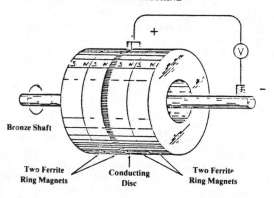

Bronze Shaft

Two Ferrite
Ring Magnets

Conducting
Disc

Two Ferrite
Ring Magnets

Tom Valone's Prototype Work

VII. *MOTOR GENERATOR UNITS & SYSTEMS*

The Kromrey electromagnetic converter has reportedly reached an efficiency of about 120%. An increase in current flow occurs under short-circuit conditions with no overheating evident. One Kromrey prototype delivered about 700 watts at a speed range of between 600 to 1200 rpm, which is generally slow for these types of m/g units. Larger units had been planned for five to twenty-five KW range which would have been ideal for home power applications.

One prominent researcher states that the original Kromrey specs-have been altered so that the unit in inoperative, as described.

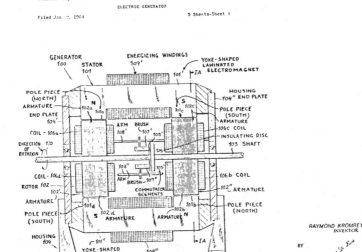

Fig.1

b) *Lawrence Jamison*   Jamison Energizer System   (Verona, Mississippi)

The Jamison Energizer System of 1980-84, is one of several prominent examples of applied tachyon field energy systems, of the several motor-generator arrangements described in this Section.

Although the full details of this high speed motor-generator system were never disclosed, it is known that it is in the motor-generator-battery class, with a very large diode necessary to control the high wattage being generated by the generator. There is no question about this system being valid and operational since a video tape demonstration was provided at the Energy Symposium at Atlanta in 1982. This operating demonstration disclosed the high noise level produced by such a high speed motor/generator system, but clearly showed a fully functional power source.

While this m/g system depends on using a standard automotive battery input, they run at over 100% efficiency with the battery being recharged as they operate under normal load conditions. The striking similarities between the Jamison Energizer System, the Gulley, Stoneburg and Watson systems leaves no doubt as to the basic operability of them, making them all active candidates for further development effort. While Mr. Jamison, (now deceased) claimed that his system was unique at the time, we now know that this was not completely true, although some of his specific components may have been custom made and proprietary, the basic principle of it is now well understood and confirmed.

The Jamison Energizer System was installed and operated in a vehicle (1977 Ford Courier pickup truck), but no operational data is available on its performance.

c) *John Gulley - Motor/Generators*   Gratz, Kentucky

The motor/generator work of John Gulley became the subject of newspaper items in the local Louisville-Courier during the late 1950's and 1960's. During this period he produced a number of operating motor-generator sets which were installed in various type of vehicles.

As a motor/generator specialist while in the U.S. Army, Mr. Gulley carried this knowledge into civilian life, where he developed various types and configurations of motor-generator combinations. These were of the basic battery-recharge type, similar to that of Jamison and Stoneburg.

VII. MOTOR/GENERATOR UNITS

The exact details of his components were never disclosed, but it is known that he did rewind the fields and armatures of both D.D. motors and generators. It is known that over-unity output operation can usually be achieved by splitting higher-than-normal voltages to satisfy both the load and battery-recharge requirements.

In personal interviews, Gulley did state that his motors were based on the solenoid principle, similar to that of Bob Teal's unit, 7d.

John Gulley demonstrated his various automotive power systems and it was reported that some commercial interest was shown for his efforts, but no further information has been revealed on his present status. There have been some comments made that Gulley was handicapped by not being able to explain the scientific theory and basis for the operation of his motor-generators, which is often the case for the hand-on type of garage-based researcher. Another unfortunate situation, was Mr. Gulley's tendency to choose exotic and sometimes outlandish names for the vehicles equipped with his special m/g sets, which did not help his cause with prospective investors. *Researchers should keep this point in mind: always maintain a conservative and explainable scientific position when providing demonstration of new energy devices!* Once scientific interviewer of Gulley's prototype became skeptical of the work after a demonstration, because of these deficiencies.

d) Bob Teal
St. Cloud, Florida (1976)

The "Magnepulsion" Motor of 1976, is a unique type of pulsed E/M motor unit which consists of multiple solenoids which are crank connected to a central drive shaft and flywheel arrangement.

The combined electromagnetic design is described in U. S. patents 4,093,880 and 4,024,421, as a magnetically operated power plant comprised of a rotary crankshaft which is rotated by means of connecting rods, pivoted to the sliding cores of electromagnets-(solenoids), as the key actuating component(s) of the unit.

Electrical current is provided to the electromagnet windings by distributor switches which are successively actuated by multiple cams on a timed camshaft. The switches receive pulses of current in timed relationship, so that solenoid thrusts are continuously and uniformly applied to the central crankshaft.

Although this type of energy conservation unit is not a true:-"free energy" unit, it does represent one of the better and basically simple energy-saving motor (such as the EvGray type) due to the aid of a high positive inertial factor provided by the main flywheel(s) on the crankshaft. As was stated elsewhere in this Manual, flywheel mass is a basically cheap means to support over-unity or energy conservation operation for any of these motor units.

Because of its simplicity, the "Magnepulsion" motor has the possibility to become an effective "over-unity" or free-energy motor when combined with some other type of solid state amplifier device,-or permanent magnet motor (Muller type Unit).

"MAGNEPULSION   MOTOR"
(Magnetically Operable Engine/-or Power Plant)
U.S. Pat. No. 4,024,421

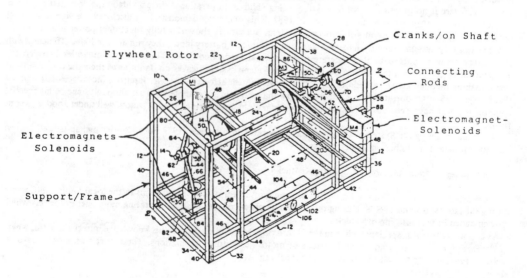

HOW THE EQUIVALENCE ENGINE WORKS

by Bruce E. DePalma

TIME ENERGY: In the conception of an idea such as a free energy machine the idea appears first as an inspiration or dream emerging as a result of incredible tension. The tension is resolved first as an intuition or dream of a machine (the language of the experimenter) which is then instituted or later explained on the basis of paradigms which form the structure of the 'language', i.e. physics. What we get is always in our genes, we just figure out how to say it. The tracks of man. We make up our own language as we go along.

A good way of figuring out or understanding where the energy comes from is to consider the Universe as a living form with both temporal and spatial extension. We can 'see' the spatial extension all around us but we only sense the time extension in the form of memory or vibrations around an object (the aura).

Although the physical body is anchored to the present the mind travels in time and space. The now consists of the point where future time intercepts past existence. This point is where an energy transfer takes place from the time energy of space to the real time of our past behavior or history. History is dead time. It exists in the mind of the rememberer until it dies and disappears. Records can be made, but they eventually disappear too. So the mind of man is finite and consists of all that he remembers or has records of. What I'm getting at is man and his world sits still in the present and the time-energy flowing through him energizes his present. So man and the world of man are like a projection slide through which the time energy plays and projects our existence out on the tapestry of created 'our time'. So man starts out as a pattern, illuminated by the time energy, he plays out his life in the emerging scroll of history. We do not go into the future. The so-called future comes into existence at the intersection of the time-energy flow with the matter (man, animals, the world) which is present. So the species has a total existence of what it remembers, together with the potentiality of what is present being energized by the time-energy flow through it.

We can say that things 'age' because of this time-energy flow. The miracle of life is that it has converted 'age' into a cycle through the mechanism of the seed. In this context, 'immortality' consists of remembering your past, up to the limit of the species itself—as manifested in any given region of space. We are all different parts in the body of the Universe, we just didn't know we were.

This blood plasma, this time energy flowing through all of us and everything, this is the fundamental energy of the Universe which animates all things. If everything which moves, and age is a form of motion, is animated by the same time energy; all lesser forms of energy, i.e. electricity, heat and pressure, are derived from that basic flow. It should be simple for us to tap this energy for the key is time is motion.

THE EQUIVALENCE ENGINE: The equivalence engine consists of a torsion pendulum excited at the nodal point. Power is taken from the end in motion. With reference to figure (1a) we see a system which is a combination of two metal discs separated and connected by an elastic member, a rod joining the discs by their centers affixed

MOST POWERFUL VERSION OF MECHANICALLY DRIVEN EQUIVALENCE ENGINE

Figure (3)

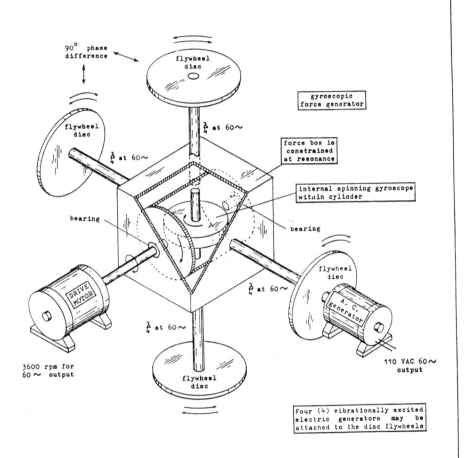

THE PHILADELPHIA ELECTRIC, equivalence engine with
torque drive by driven gyroscopic precession

drawn: ECD 6-77

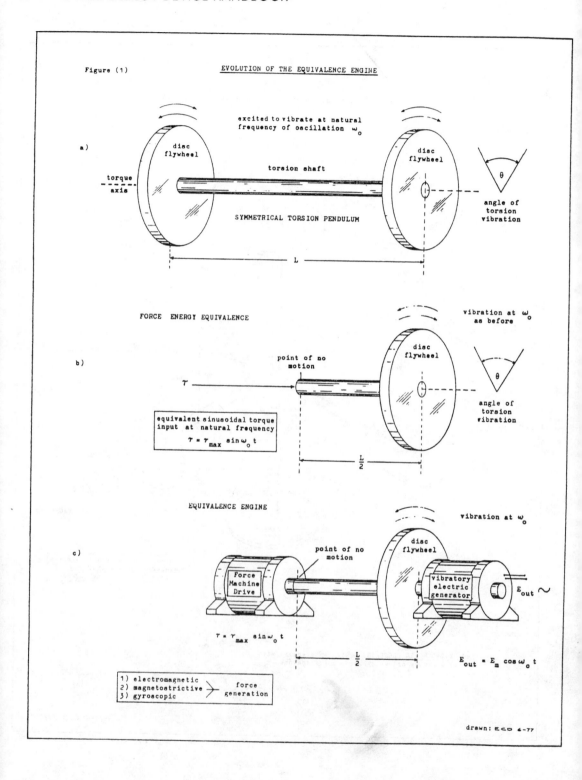

Figure (1)   EVOLUTION OF THE EQUIVALENCE ENGINE

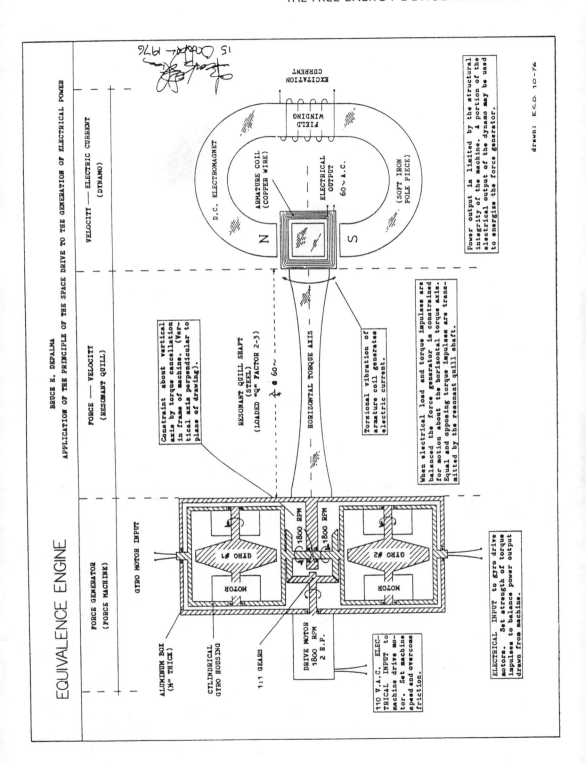

EQUIVALENCE ENGINE

BRUCE E. DEPALMA

APPLICATION OF THE PRINCIPLE OF THE SPACE DRIVE TO THE GENERATION OF ELECTRICAL POWER

FORCE GENERATOR
(FORCE MACHINE)

FORCE — VELOCITY
(RESONANT QUILL)

VELOCITY — ELECTRIC CURRENT
(DYNAMO)

GYRO MOTOR INPUT

ALUMINUM BOX
(¼" THICK)

CYLINDRICAL
GYRO HOUSING

1:1 GEARS

110 V.A.C. ELEC-
TRICAL INPUT to
machine drive mo-
tor. Set machine
speed and overcome
friction.

DRIVE MOTOR
1800 RPM
2 H.P.

ELECTRICAL INPUT to gyro drive
motors. Set strength of torque
impulses to balance power output
drawn from machine.

GYRO #1   1800 RPM
MOTOR

GYRO #2   1800 RPM
MOTOR

Constraint about vertical
axis by torque cancellation
in frame of machine. (Vertical axis perpendicular to
plane of drawing).

RESONANT QUILL SHAFT
(STEEL)
(LOADED "Q" FACTOR 2-3)

↕ @ 60~

HORIZONTAL TORQUE AXIS

Torsional vibration of
armature coil generates
electric current.

When electrical load and torque impulses are
balanced the force generator is constrained
for motion about the horizontal torque axis.
Equal and opposing torque impulses are transmitted by the resonant quill shaft.

D.C. ELECTROMAGNET

N   S

ARMATURE COIL
(COPPER WIRE)

ELECTRICAL
OUTPUT
60~ A.C.

(SOFT IRON
POLE PIECE)

FIELD
WINDING

EXCITATION
CURRENT

Power output is limited by the structural
integrity of the machine. A portion of the
electrical output of the dynamo may be used
to energize the force generator.

drawn: E.C.D. 10-76

15 October 1976

8

Phenomenon of Electric Charge Generation by Space Rotation

Phenomenon Of Electric Charge Generation By Space Rotation

Paramahamsa Tewari
Chief Project Engineer
Kaiga Atomic Power Project
Nuclear Power Corporation
KARWAR, INDIA

Abstract

The medium of space (absolute vacuum without matter) is defined as an incompressible, zero-mass, nonviscous, continuous and mobile entity which, in its rotation at the limiting speed of light as a submicro vortex, creates electrons. The property of electric charge of electron and its electrostatic field can be shown to be the effect of rotation of space around the electron's centre. The mass property of electron is seen to be arising due to the creation of a fieldless spherical void (hole) at electron's centre where space rotates at the limiting speed of light.

From the "rotating-space" and "void-centre" model of an electron, new equations on the electron's mass and charge are obtained. An experimental proof to this new philosophy of matter as a "dynamic condition of space" is had in the generation of electrical energy from the inter-atomic space of a rotating electromagnet. A cylindrical electromagnet rotated about its axis develops voltage between the axis and the cylindrical surface of rotation. By drawing electrical power through suitable liquid metal brushes placed at the axis and the cylindrical surface of rotation it can be shown that a rise in electrical power output does not fully reflect in the equivalent increase in the mechanical input used to rotate the device. The incremental ratio is seen to be more than unity in numerous experiments carried out on space power generators in recent past.

Introduction

Scientific debates on the nature of space (absolute vacuum) around the start of this century took a general view that space serves to transmit fields (electromagnetic, gravity) and, beyond that, it has no independent existence of own as an entity that can generate energy in some form which enables it to be termed as a "real" physical entity. In other words, the generally accepted position then taken was that space is an empty extension of "nothingness" without any physical attributes in which fields are transmitted. Many refutations to the above view both experimental and theoretical, and claims on the existence of either, that is, space with physical attributes, could not provide an alternate theory so as to change the widely accepted position of nonphysical nature of space.

A search for the most basic and universal source of fields of matter in general, and energy and matter in particular, has led to the formulation (by the writer) of space vortex theory (SVT) in which physical space has been postulated.

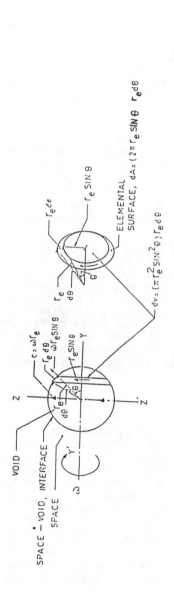

ω : ANGULAR VELOCITY OF INTERFACE ALONG , Y-Y'.

VOID: FIELDLE: SPHERICAL HOLE IN SPACE.

SPACE: NON-VISCOUS, MOBILE, CONTINUOUS, INCOMPRESSIBLE.

VOID-RADIUS $r_e \simeq 10^{-13}$ CM

CHARGE ON ELEMENTAL RING SURFACE, dq = RING AREA X SPEED OF CIRCULATING SPACE ON RING SURFACE

$$dq = dA \,(\omega r_e \sin\theta)$$

ELECTRONIC CHARGE, $q_e = \int_0^\pi (2\pi r_e \sin\theta \, r_e \, d\theta)(\omega r_e \sin\theta)$

$$q_e = (\pi/4)(4\pi r_e^2 c)$$

DIMENSIONS OF q_e = LENGTH^3 / TIME

REST-MASS OF ELEMENTAL DISC OF VOID.

dm = dv X SPEED OF CIRCULATING SPACE AT THE INTERFACE OF THE ELEMENT.

$$dm = (\pi r_e^2 \sin^2\theta \, r_e \, d\theta)\, \omega r_e \sin\theta$$

ELECTRONIC REST-MASS, $m_e = \int_0^\pi \pi c\, r_e^3 \sin^3\theta \, d\theta = (4\pi/3) r_e^3 c$

DIMENSIONS OF m_e = LENGTH^4 / TIME

VOID CENTRE OF ELECTRON

FIGURE - I

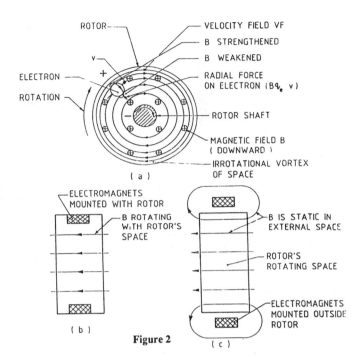

Figure 2

Basics of Space Vortex Theory (SVT)

The absolute vacuum in SVT is a nonmaterial entity--an uncompressible, nonviscous, massless and continuous medium and its dynamical field equations or laws are formulated introducing concept of energy-fields. An electron is postulated to be an irrotational vortex of space with a spherical central void of sub-microscopic radius of about ten raised to the power minus eleven centimeter, within which the space with physical attributes develops discontinuity of energy field (Fig.1). New equations that quantify mass and charge of electrons and enable the computation of energy for its creation and annihilation are derived in SVT through an analysis based on the dynamical equations. It also follows from the theory that the electron is the only fundamental particle that can build all of the other stable particles of matter of the universe.

Experimental Proof

The above structure of the electron is substantiated through some recent experiments on electromagnetic induction. These experiments have brought to light a new phenomenon in which electromotive force is generated in an electrical conductor even if there is no relative motion between the conductor and the magnetic field which is unlike the principle behind the operation of modern electrical generators for which relative motion between the conductor and the magnetic field is a must (Fig.2). Further, the electrical power generation in the newly discovered phenomenon does not fully reflect in a proportional increase of the input power, thereby giving an efficiency of more than unity, and revealing a new vital fact that absolute vacuum in the interatomic space of matter can generate electrical charge and power when it is set in a dynamic state (Fig. 3). An experiment similar to above was first

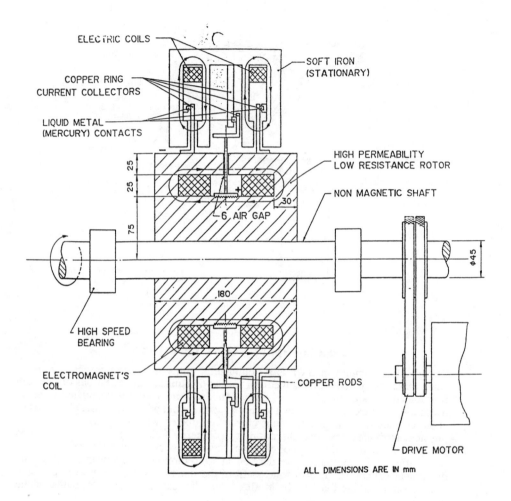

ALL DIMENSIONS ARE IN mm

Figure 4. Space power perpetual machine.

The present industrial culture demands a continually increasing growth of electrical power. Since the limited resources of thermal, hydroelectrical and nuclear power cannot last for centuries, the need that has now become vital is to discover a power source which is independent of all material resources, so that a regulated generation of power from the presently known technologies is supplemented to a good extent from this eternal source.

A technological breakthrough, which will enable generation of energy directly from space, is in the offing. Nature has, however, already achieved this conversion from space energy to matter in the generation of the cosmic matter of the universe.

done by Michael Faraday in the year 1830. He discovered the above phenomenon as regards the production of electromotive force even when the relative motion between the magnetic field and the conductor was zero. He, however, does not appear to have made measurements on the incremental power ratio from such a device. The writer first learned of this effect from Bruce De Palma's letters to him (1978-79).

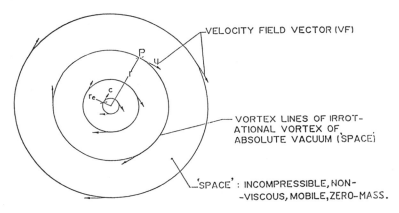

AT ANY POINT P OF A VORTEX LINE, ur= CONSTANT,
WHEN r = r_o, u = c.
THEREFORE, cr_o = CONSTANT,
AND u = cr_o/r.

Figure 3. Irrotational vortex of absolute vacuum (space) -- two dimensional.

The quantity of electric charge generated by the rotation of the interatomic space in a rotating iron disc requires the use of equations of SVT on mass and charge of electron. The incremental output power is shown to be directly proportional to the volume of space within the rotor of space power generator and the speed of rotation of the rotor (Fig.4). The voltage generated is at a low range (expected to be 6 volts DC with about a 14 inch diameter rotor) but the current is of high magnitude (in thousands of amperes when liquid metal is used for current collection).

In one of the recent experiments on a space power generator (Fig. 4), an incremental power ratio of 4.5 has been obtained.

Conclusion

Building of space power generators of about 100 KW size should be possible to be achieved with conventional technological methods in the near future.

In addition to the discovery, of a novel technology for power generation, the following crucial change in the scientific and philosophical outlook that would be brought about with the new phenomenon of space power generation are:

[1] The fundamental state of universal energy is eternally latent in the dynamic space of the universe.

[2] The cosmic matter is generated from the energy of space in repeated cycles of creation and annihilation.

[3] Space is not an empty extension but rather the most fundamental entity which alone gives reality to the cosmic world.

9

The Worthington Magnet Motor

WORTHINGTON MAGNETIC MOTOR

Our first look into the Permanent Magnet Motor is this one shown here from the year 1929 invented and patented by H.L. Worthington.

Worthington's idea was to use a rotor consisting of permanent magnets, and a surrounding field also consisting of permanent magnets. The rotor magnets and the field magnets were to have like poles that would face each other at all times causing a repulsive magnetic force to push each magnet away from each other. By placing magnets into a position such as that shown in the drawing, with eight field magnets and four rotor magnets, the push from each magnet equals the push of all others and therefore the device would not rotate.

However, Worthington installed a set of two electric solenoids and a rocker arm at each set of two field magnets. Each rocker arm was to have a heavy iron roller attached to each end and was to be positioned so as to move the rollers into and out of the magnetic field of one magnet at strategic moments of rotation to cause the rotor magnets to be drawn toward the field magnets until they would pass a particular point wherein the roller would change positions,causing the rotor to rotate because of magnetic repulsion.

Each roller was to act as a shunt causing two north poles or two south poles to attract each other until the magnets were in position to continue rotation because of repulsion when the shunt was removed from the position that caused the attraction. Worthington's idea was to use electric solenoids to cause the rocker arm to insert and remove the shunt rollers from the magnetic field at strategic moments of the rotation, thus causing continued rotation.

The drawing shows what looks like bar magnets, but in reality his design used horseshoe magnets, and the drawing would not show south poles as they would also all be facing each other and would be hidden directly behind the north poles as seen in the particular view of the drawing. However, if the device were to actually work, either type of permanent magnet would suffice.

The one thing wrong with Worthington's theory of placing a shunt in the magnetic force field of two like poles to cause them to attract each other, is that placing a shunt between two like poles DOES NOT cause them to attract each other. They will each attract the shunt, but not each other — and if the shunt is of soft iron the two like poles will in fact move toward each other, but only once they are close to the shunt — close enough for the magnetic force of attraction toward the shunt to be stronger than the magnetic force of repulsion of the two like poles of the permanent magnets.

In my own experiments with permanent magnets I have found that you can cause two like poles to move toward each other, but again, only after they are close enough for the attraction to the shunt to be much stronger than the repulsion force. Up until that point the two like poles will repel each other. I have not yet been able to get two like poles to move toward each other to a point where upon removal of the shunt they would repel each other and keep the moving magnet going in the same direction rather than to reverse directions going back the same way it came from to get into the field.

Aug. 13, 1929.   H. L. WORTHINGTON   1,724,446
MAGNETIC MOTOR

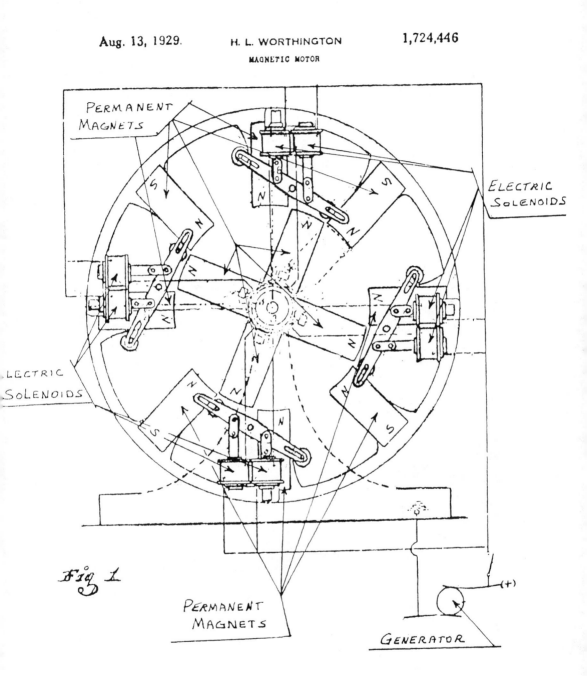

PERMANENT MAGNETS

ELECTRIC SOLENOIDS

LECTRIC SOLENOIDS

Fig 1

PERMANENT MAGNETS

GENERATOR

(+)

INVENTOR

Harry L. Worthington.

BY

R. Freeman Vale,

ATTORNEY

MAGNETIC MOTOR MAY HERALD QUANTUM SHIFT

Troy Reed is the inventor of a new magnetic generator which may be in production and available to the public soon. The project, which has been in development since Reed's initial vision in 1959, has culminated in a magnetic generator which can provide all the power necessary for a small home. But that, he contends, is just the beginning.

With further development, the potentials are limitless. They include everything from powering automobiles to large industry without any outside source of energy—no fossil fuels, no nuclear, no wind, sun, or water, just magnets.

Reed is a paradigm buster. Unfettered by a conventional scientific background, he was unhampered by basic scientific dogma such as the second law of thermodynamics, "you can't have power output without energy input." Unfamiliar with "natural limits" as defined by science, he simply stepped beyond them. He likes to tell the story of the Wright brothers who invented the first airplane in the USA. "The US Patent Office refused their first application, all because the experts said it was impossible for anything heavier than air to fly!"

The magnetic motor began in 1959 as an eerie image floating before his eyes. Troy was bent over a project at the Tulsa machine shop where he worked, grumbling about laboring so hard for a mere 70 cents an hour. "There's got to be something better." With this thought, the image appeared briefly and then disappeared, only to return about six months later. It was an object with wheels and cylinders.

"Six or seven weeks went by. I walked up to my machine, which I had cleaned the night before. Two magnets lay there with shavings on them. Where they came from, I'll never know. The mysterious magnets and the vision haunted me for two years until I decided I had to do something about it," says Reed.

That began his odyssey into the surrealistic world of magnetism and its potential for being harnessed into a dependable and perpetual source of free energy.

Reed's odyssey reached a turning point in 1991 (when) he produced the first working model of the motor...This prototype was nearly seven feet tall and weighed over 500 pounds. It could drive a conventional electric generator to output 500 watts at 67 volts and produced enough energy to run power tools and household appliances. More importantly it proved that the principle worked. It was a functional prototype which was examined and verified by independent physicists and engineers.

The current sophisticated high-tech model weighs less than 200 pounds and is about the size of a large hat box. It can drive a 7-kilowatt electric generator, producing enough energy to power a small home.

The Reed Magnetic Motor is non-polluting and environmentally safe. Its four-cylinder injection engine runs on power generated by the repulsion force of magnets. This is made possible by the fact that magnetic force fields can be deformed, reshaped, rechannelled and manipulated in various conventional ways.

The original model was extremely simple and consists of only four parts—a crank shaft, injectors, disks, and magnets.

The crank shaft has four rods which are interconnected to the injectors. The cylinders merely house the injector rods which connect the crank shaft to the injectors.

The injectors are spring loaded and act like a ballpoint pen—they cock, push and release. This release of the four injectors does two things—it regulates the motor's RPM (revolutions per minute) and it makes the motor run smoothly.

There are two rotating disks at each end of the crank shaft. These carry 16 magnets each. Built into each end of the engine housing are more disks which hold additional magnets, some which move and some which are stationary.

The magnets that can be used to work in the motor range from a magnetic power rating of 3.5 (ceramic magnets) to a #40 neodymium. Of course, the stronger the magnet, the more horsepower, braking horsepower and torque the engine has.

The magnetic motor produces a direct mechanical effect. By manipulating the shape of magnetic fields, it turns a drive shaft to produce rotary power. Magnetic fields within the unit are continuously reshaped and rechannelled in a synchronized manner. These ever-changing fields manipulate the net strength and direction of the magnets' forces, driving the unit's rotary shaft at thousands of revolutions per minute.

Mr. Reed's invention is unlike any previous magnetic motors because it uses both rotating and stationary permanent magnet disks, along with spring-driven "injectors." These provide the necessary intermittent force to convert the magnetic reactions into smooth continuous rotations.

Principles governing the motor's operation are similar to the laws of electromagnetic attraction and repulsion, Mr. Reed says.

Two disks are located on each end of the motor's housing. The outer disks are stationary, while the two inner ones, which are mounted on a common shaft, are designed to turn freely.

Eight permanent magnets are attached, equally spaced, to the outer rims of each of the four disks.

Another major principle involves spring-type injectors. These operate very much like the push-pull mechanism of a common ballpoint pen. The injectors provide a "kick" to the crank shaft, which helps to propel the shaft forward and takes advantage of the repulsion forces created by the next series of magnets.

The design of the motor allows it to be scaled up to any size for a variety of applications. Because of its simplicity, few operating parts, and low operating speed—which results in little wear and tear—the motor has proven both dependable and durable in tests.

Because of this simplicity, Mr. Reed predicts that the cost of production and parts, excluding the magnets, will be low.

Some visionary thinkers have said for years that, as a species, we are on the verge of a quantum shift. There are different ideas about what may trigger this shift and how it might take place.

These ideas include massive Earth changes which totally disrupt "business as usual" and force a much reduced human population into new priorities and ways of living cooperatively and peacefully.

Others anticipate what they call an Ascension of Humanity, which ranges from shifting into higher dimensions of consciousness, all the way up to a "beam me up now, Scottie" scenario.

Only time will tell what unfolds in current consensus reality. One thing seems sure however.

A new magnetic motor like this may precipitate changes, within the next two decades, which were fanciful science fiction a few years ago.

An invention like Mr. Reed's could well bring about a major shift in economies and the power of money.

Freedom to live where and how we choose unlocks untold financial resources from the industrial and political power brokers who control the production and distribution of energy.

Such an idea may be the telling factor which will create a quantum shift, pushing us gently but firmly into a much larger understanding of what is real.

KELLY MAGNETIC WHEEL DRIVE

The 1979 design of D.A. Kelly differs in several respects from the two previous designs. Although the wheel that provides torque from the motor is almost identical with its permanent magnets being faced in the same direction (with all like poles facing outward), a major difference occurs in the motor's driving force.

Instead of having sets of iron-based shunts positioned into and out of the magnetic force field to alternately attract and repel the magnets of the drive wheel, the Kelly design uses no shunts at all. Instead it uses oscillating rocker arms to insert and remove field magnets themselves into and out of the magnetic force field of the drive wheel magnets. To accomplish the oscillating action of the arms the designer uses electric motors and eccentrics plus appropriate linkage to all sets of rocker arms.

Each rocker arm is fitted with two magnets, one with like pole to be facing like poles of the drive wheel magnets and the other with unlike pole to be facing the drive wheel magnets. Kelly says in his patent that this causes a pull and then a push upon each drive wheel magnet as the rocker arm oscillates from one position to the next.

The Kelly patent states that the use of solar cells and battery storage is to be the primary source of current to power the motors driving the oscillating rocker arms and respective linkage, yet the patent also states that the torque provided by the drive wheel of the permanent magnet motor is considerably more than the torque required to operate the oscillating rocker arms and linkage.

If the Kelly design does in fact work as claimed it would be simple to have the drive wheel connected to an alternator which would provide the necessary voltage to the electric motors, thus making it a self-propelled motor just as is claimed for the Tracy and Derouin design and patent. Also as one may wonder about these devices, why even try to design a permanent magnet motor if it requires an outside source of energy such as electricity—unless the device itself can provide the necessary current to continue its operation.

There are A.C. current motors already available on the open market that approach 100% efficiency from the current they draw. Such motors are not cheap to purchase, but they are available now—so why build a permanent magnet motor unless it can power itself without the need for current from any other source than itself?

United States Patent

Kelly

4,179,633

Dec. 18, 1979

OSCILLATING ROCKER ARMS (8)

ELECTRIC MOTOR

ALL OSCILLATING ROCKER ARMS ARE INTERCONNECTED BY LINKAGE NOT SHOWN

ELECTRIC MOTOR

MAGNETIC WHEEL DRIVE

Inventor: Donald A. Kelly

KINNISON PATENT

The 1975 patent by R.W. Kinnison, as with many others, uses a form of shunt to effect a counteraction to the normal repulsion of two like poles of permanent magnets. However, Kinnison calls his use of the magnetic material (such as iron) being inserted and removed from the magnetic field a type of "diverter."

Kinnison uses horseshoe magnets placed at strategic positions for a set of stationary magnets, and another horseshoe magnet attached to a rotatable shaft. His patent claims that an electric solenoid can be used to move the diverter into and out of the magnetic field, or one of several different mechanical means can be used, such as a gearing system or the use of an eccentric or cam.

Kinnison claims that the motor works only on the repulsion force of the permanent magnets, stating that the diverters only divert the magnetic force of two like poles' repulsion for each other. He says that the diverters are made of non-ferrous metal (non-magnetic) with inserts of ferrous (magnetic) metal at strategic points.

As the rotating magnet moves into position where two like poles would normally repel each other the diverter is moved into position to cause a diverting of the magnetic force of repulsion, thus allowing the like pole of the rotating magnet to move fully into the field of the stationary like pole. Once into position the diverter is removed causing a normal repulsion of the two like poles and thus the rotating magnet continues to rotate.

With two north poles and two south poles in the stationary form of the device and one north pole and one south pole on a rotatable shaft, the device would produce four acts of repulsion per revolution and each rotating pole would pass through a position of neutrality four times per revolution. It must be noted also that since each stationary and each rotating magnet will have a north and a south pole facing toward another, the stationary magnets must be strategically positioned so as the unlike poles will not affect operation.

United States Patent

Kinnison

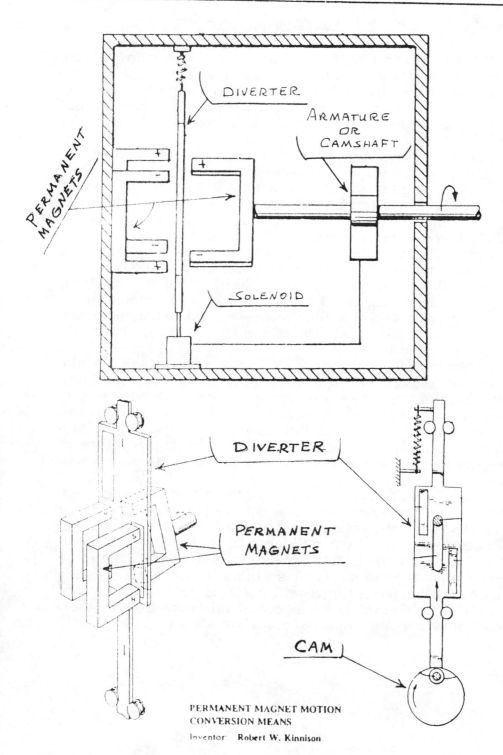

DIVERTER

ARMATURE OR CAMSHAFT

PERMANENT MAGNETS

SOLENOID

DIVERTER

PERMANENT MAGNETS

CAM

PERMANENT MAGNET MOTION
CONVERSION MEANS

Inventor: Robert W. Kinnison

JINES AND JINES PATENT

The 1969 patent of Jines and Jines is different from other types of permanent magnet motors in the respect that it uses only magnetic forces of attraction, disregarding repulsive forces altogether. It also uses a highly magnetic part as the driven portion of the motor instead of a permanent magnet.

The top drawing shows two stators and a rotor. Each stator has permanent magnets attached in a stationary manner with each having like poles facing the rotor. The rotor consists of a highly magnetic piece (such as soft iron) attached to one side and a non-magnetic counterweight at 180 degrees opposite, and two cams each placed on opposite sides of the rotor shaft.

With the stators and their respective stationary magnets turned to face each other and turned at an angle so as two magnets are not directly opposite each other but some degrees off, and the rotor to be positioned between the stators—each magnet will attract the part of the rotor made of soft iron. The repelling force of the like pole facing magnets has nothing to do with causing the rotor to rotate.

What the Jineses claim is that a permanent magnet can be made to have no magnetic effect on the rotor until strategic moments of its rotation to keep it in motion. They claim this can be done by having the magnets entirely encased in a magnetic metal such as soft iron.

The center drawing shows a view of either of the two stators with its stationary magnets. The stator is made of soft iron and encases all sides of each magnet except the one facing the rotor (and also the other stator). Installed into the stator is a set of what the Jineses call magnetic shields to be activated by the cams upon one side of the rotor shaft. Each of the shields is also made of iron and when in place covering the magnets, according to the Jineses, shields the magnet so as to reduce or eliminate the magnetic effect it has on the rotor. As the rotor turns, each shield is dropped toward the shaft at strategic times when the magnetic iron piece on the rotor will be pulled toward the magnet.

As all magnets are covered except the particular one that must be uncovered at the moment the rotor magnetic piece is about to move upon its magnetic field, and the uncovered magnet is again shielded at the moment it must leave the magnetic field, the rotor continues rotation—at least according to the inventors.

The lower drawing shows the magnet as it is installed into the iron stator and the iron shield that moves to cover or uncover the remaining face of the magnet. As the rotor rotates each magnet will be uncovered, by movement of the shields caused by the actuating rods and their travel up or down the rotor cams, and each magnet in turn will attract the iron weight of the rotor. One magnet of each stator will be uncovered in succession: first one magnet of stator #1 and then one magnet of stator #2 and so on around the stators. Each magnet is to be again covered at the moment the rotor piece must leave its magnetic field and move on the uncovered magnet, in turn causing a continuous rotation of the rotor in whichever direction it is given its initial start. At least, so say the Jineses. Personally I have not found anything that a magnetic field cannot pass through.

Sept. 23, 1969   J. E. JINES ET AL   3,469,130
           MEANS FOR SHIELDING AND UNSHIELDING PERMANENT MAGNETS
                AND MAGNETIC MOTORS UTILIZING SAME

215

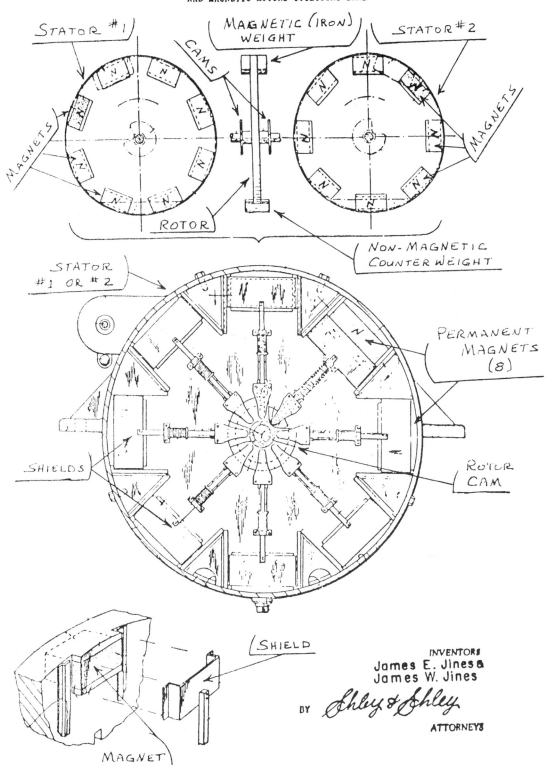

STATOR #1

MAGNETIC (IRON) WEIGHT

STATOR #2

CAMS

MAGNETS

MAGNETS

ROTOR

NON-MAGNETIC COUNTER WEIGHT

STATOR #1 OR #2

PERMANENT MAGNETS (8)

SHIELDS

ROTOR CAM

SHIELD

MAGNET

INVENTORS
James E. Jines &
James W. Jines

BY Shley & Shley

ATTORNEYS

PUTT ENERGY CONVERSION SYSTEM

The 1976 patent of J.W. Putt shown here is not, as such, one of the many permanent magnet motors that have been designed and patented in this century. But it is presented here because it does in fact use permanent magnets, and according to the patent specifications, can be used as the power source for an automobile.

The device is a form of hydraulic pump. Figure two shows a set of magnets, ten in all, attached to piston-type hydraulic pumps and placed around a circular housing. Within the housing is a rotor of permanent magnets, and when this rotor is driven the magnetic force of attraction and repulsion against the surrounding magnets causes them to pump hydraulic fluid—in this particular case, according to the patent specification, at 1000 psi and 2400 cubic inches per minute. Yet the rotor must be driven by an outside source of power.

Figure one shows the device in configuration for the drive train of an automobile. Putt says in the patent specs that the device will produce enough hydraulic power to drive an automobile with the use of an ordinary hydraulic motor designed for such use (hydraulic drive motors are available on the open market).

Putt also says that the magnetic rotor is to be driven by the same means, a hydraulic motor. Putt specifies, more than once, in his patent specs that the device is designed with automotive drive as a use for the device. And since an automobile does not ordinarily carry on board any kind of source for hydraulic pressure and volume in enough capacity to drive the vehicle, it must be assumed that the closed hydraulic system pressure and volume must not only drive the vehicle but also drive the magnetic rotor.

The patent specs do not divulge the source of hydraulic power that drives the hydraulic motor driving the magnetic rotor. But since it would be ridiculous to design an automotive drive train that required an engine, such as a gasoline or diesel engine, to drive a hydraulic pump to drive a hydraulic motor that drives a magnetic rotor of the magnetic hydraulic pump to drive another hydraulic motor to ultimately drive the vehicle—again it must be assumed that the permanent magnet hydraulic pump must provide enough volume and pressure to drive both itself and the vehicle. Otherwise why spend the money on so many, many costly components and still end up using petroleum fuels?

The Putt patent also specifies that other means of power can also be derived from the device instead of hydraulic power—such as electrical generation through the use of linear generators at each of the surrounding field magnets, or by mechanical means from the up and down or back and forth motion of the field magnets as the rotor causes them to move upon its rotation.

However, the hydraulic type of power is gone into in detailed specs in the patent text, and is shown somewhat in these drawings. Putt's text says that the power required to drive the magnetic rotor is very small in comparison to the power output of the device because each of the rotor magnets moves into opposite magnetic fields at the same time with one moving into a magnetic attraction force at the same time as the other moves into a magnetic repulsion force, thus almost equalizing the force required to drive the rotor with the only power being required as the difference between attraction and repulsion.

FIG.1

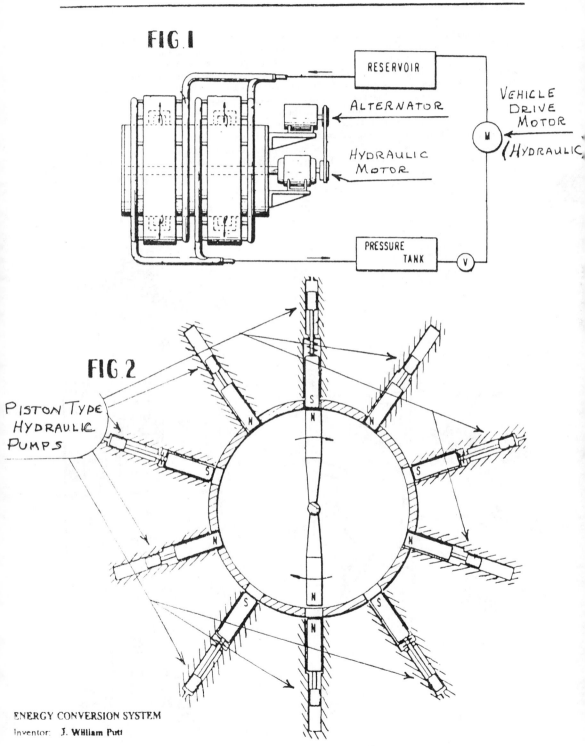

RESERVOIR

ALTERNATOR

HYDRAULIC MOTOR

VEHICLE DRIVE MOTOR

M

(HYDRAULIC

PRESSURE TANK

V

FIG.2

PISTON TYPE HYDRAULIC PUMPS

ENERGY CONVERSION SYSTEM

Inventor: J. William Putt

J.W. Putt   PM/EM Motor-U.S. Patent No. 3,992,132   (Nov. 1976)

The permanent magnet/-electromagnetic motor of J. W. Putt is based on converting primary motion in a given path to produce secondary motion in a transverse direction to the given path, as illustrated.

This ununual type of PM/EM energy conversion unit which utilizes radial displacement of permanent magnets to drive pistons for a pumping device should be considered as a hybrid type of arrangement, as noted.

The radial pistons are used as a hydraulic pumping means and are connected together to provide an output of pressurized fluid which may be stored under pressure for convenient usage. The fluid may be used to drive a hydraulic motor and then recirculated through a reservoir to the individual pump means for each of the secondary magnets.

The basic invention involves the balancing of the magnetic forces of attraction and repulsion. This is achieved by having a plurality of interconnected primary magnets with polarities which coact with the polarities of the transversely movable secondary magnets so that the force of magnetic attraction in one direction parallel to the given path of relative movement are substantially equal to the forces of magnetic repulsion in the opposite direction parallel to the given path. This arrangement results in a minimization of th energy required to produce relative movement in the given path between the primary and secondary magnets.

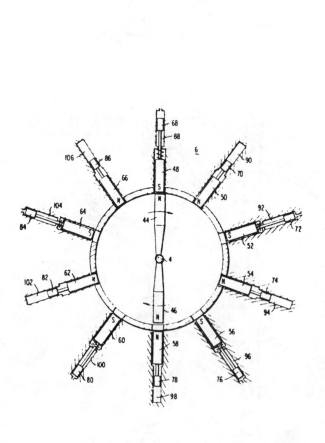

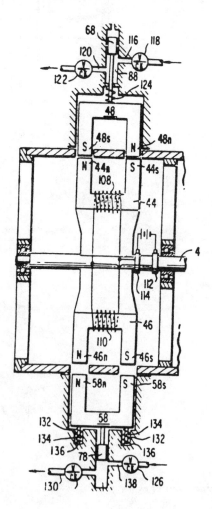

TRACY et al. PATENT

With the 1972 patent of Tracy and Derouin we get a more recent look at the permanent magnet motor. This design holds with the same idea of placing a shunt at strategic points and at strategic moments of motor operation as does the previous Worthington design. However, a different approach is taken in that, instead of having the shunt roll into place alongside and touching the stationary magnet as with the Worthington design, the Tracy and Derouin design has the shunt moving in and out of the magnetic force field without actually touching the magnet AND also into the field directly between the stationary and the moving magnet.

This design, as with the Worthington design, uses electric solenoids to move the shunts in and out of position—and in addition two designs are given: a rotary design similar to any ordinary electric motor, and a reciprocating design similar in fashion to any ordinary internal combustion engine. The reciprocating design would, however, seem to be a waste of energy in overcoming friction of the many, many unnecessary reciprocating parts (upper drawing) such as connecting rods, insert bearings and piston-type sliding magnets. The rotary type would be considerably more efficient if in fact the device will work as claimed by the inventors.

Again, and as with the Worthington design, the idea is to insert a shunt (preferably of iron or iron alloy) between two like poles of a set of magnets, one stationary and one moveable, to cause the moveable magnet to move toward the stationary magnet, and then at a strategic point of operation to remove the shunt causing the magnets to repel each other. To move the shunts into and out of position an electric solenoid attached to each shunt, energized and de-energized at strategic moments of operation, causes the moveable magnets to be moved either toward or be repelled away from the stationary magnets and thus cause continuous operation of the motor.

As with the Worthington design, an outside source of current is required to operate the electric solenoids. However, the Worthington patent does not say specifically that the outside source of current is to be produced by the motor itself—only that a source such as a generator could be used. The Tracy and Derouin patent does state in specific terms that the source (an alternator) is to be driven by the motor itself, that the battery (shown in the drawings) is used only for starting purposes and then regenerated by the alternator, and that the alternator (driven by the motor itself) is to supply the necessary current for normal operation, thereby causing continued operation of the motor with NO NEED FOR FUEL OF ANY KIND.

As previously stated, two like poles of two permanent magnets can be made to move toward one another by placing an iron shunt between them. This takes place because the magnetic force of attraction will be greater than the magnetic force of repulsion—AND since magnetic force of attraction IS GREATER than magnetic force of repulsion in the same magnet, these designs have not considered the amount of force required to remove the iron-based shunts from position.

Will the required energy to remove the shunts from position be of greater demand than the repulsive forces of each magnet provided for that purpose PLUS continue operation of the device itself?

United States Patent

Tracy et al.

3,703,653

Nov. 21, 1972

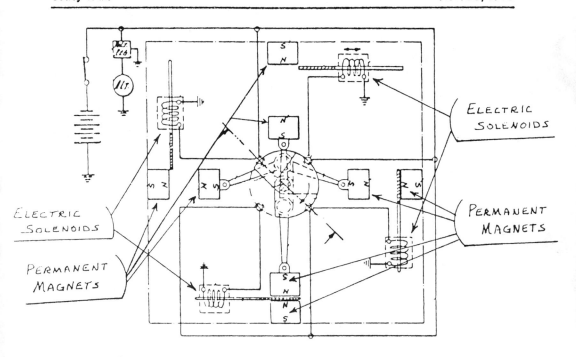

ELECTRIC SOLENOIDS

ELECTRIC SOLENOIDS

PERMANENT MAGNETS

PERMANENT MAGNETS

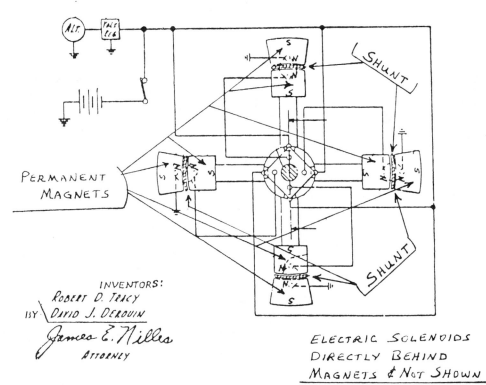

PERMANENT MAGNETS

SHUNT

SHUNT

INVENTORS:
ROBERT D. TRACY
BY DAVID J. DEROUIN

James E. Nilles
ATTORNEY

ELECTRIC SOLENOIDS
DIRECTLY BEHIND
MAGNETS & NOT SHOWN

10
The Swiss M-L Convertor

h) *Swiss M-L Converter*

The Swiss M-L converter is a fully symmetrical, influence-type energy converter, which is essentially based on the Wimhurst electrostatic generator with its twin, matched counter-rotating discs.

It is apparent that this unit design has been substantially upgraded over the old Wimhurst electrostatic generators, but still has the characteristic metallic foil sectors which both generate and carry small charges of electricity to be stored in matched capacitors. Each sector accumulates the charges derived by influence with the other sectors.

In the old Wimhurst units diagonal neutralizing brushes on each opposite disc distribute the correct charges to the sectors as they revolve, but in this new M-L converter this function is accomplished by a crystal diode at higher efficiencies than the older design.

Two collection brushes collect the accumulating charges and conduct them to the storage capacitor, located at the top of this new design. Unlike the old Wimhurst design, this new converter utilizes several new and improved features such as two horseshoe magnets with matched coils, and a hollow cylindrical magnet as a part of the diode function, and two Leyden jars or flasks, which apparently serve as the final capacitor function for the converter.

It becomes apparent that this new converter susbtantially increases the current (amperage) flow with the addition of the coil and magnets combination, as in the Coler solid-state devices. The use of top grade components, such as gold-plated contacts, control electrodes and dual capacitor stages insure much higher conversion efficiences than was possible with the old Wimhurst machines.

The general specification for the operating prototype are:

1) *Efficiency: 1:10₆,* due to self-sustained operation. The unit is started by hand revolving, with no other input power source required!!

2) *Constant Power Output: 300 volts* @ 10 Amperes = 3KW.

3) Dimensions: 43.31'' wide, 17.72'' deep, 23.62'' high

4) Weight: 44 Lbs.   Operating speed 60rpm.

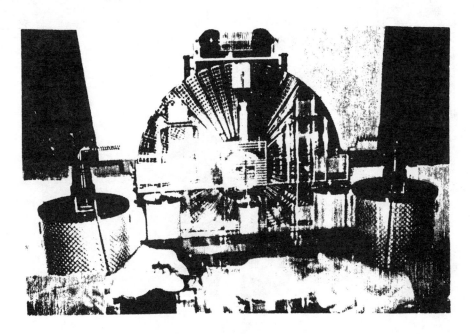

**WIMHURST
ELECTROSTATIC
GENERATOR**

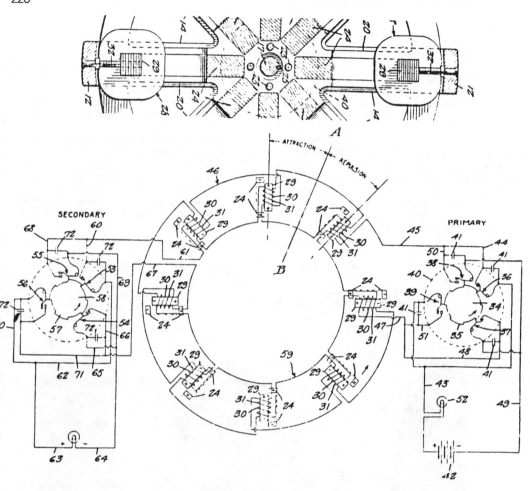

M-L CONVERTER

VII. *GENERATOR UNIT (Self-sustaining Type)*

Analysis of the Swiss M-L Converter, with multiple electrical circuits.

It is evident that this excellent o/u/o Converter unit is based on the Winshurst electrostatic generator which utilizes

It is evident that this excellent o/u/o Converter unit is based on the Wimshurst electrostatic generator which utilizes multiple *steel* segments. These Wimshurst E/S generators are made with either steel or aluminum segements, with the aluminum segments being true electrostatic elements.

When *steel* segments are used on the twin discs of this unit, the *Searl Effect* is in evidence, with E/M conversion made at the rim/periphery of the discs through passive electromagnets.

This unique o/u/o unit becomes an ideal converter since both high voltage A.C. and moderate A.C. amperage can be simultaneously generated through two separate electrical circuits from the discs. The twin disc's conventional conductive brushes pick-off the high voltage A.C., while the rim electromagnet's coils produce a useful E.M.F. (useful amperage level). When permanent horseshoe magnets with coils are utilized, as in this present Swiss unit, then the E.M.F. output is enhanced to a considerable extent, as is evident in the specs. for this M-L unit.

The self-propulsion, after hand starting, is achieved through the adaptation of the *Poggendorff principle* (a German scientist of the 1870's) in which slanted conductive brushes produce self-rotation in electrostatic motors, (not generators).

In regard to the special crystal diode module, this component most probably provides the dual functions of frequency regulation and capacitance amplifier — to the two Leyden jars. This special diode-capacitor provides frequency output regulation and capacitance amplification as part of the electrical resonance circuit, since it is connected with the horseshoe magnet coils.

This unit is essentially comprised of three separate electrical circuits, which are:

1) The high voltage A.C. output from the twin discs as a conventional Wimshurst electrostatic generator.
2) A moderate A.C. amperage circuit produced by the dual horseshow magnet coils (Searl Effect) as the plus and minus discs pass by them. (Pulsed D.C. output at 50 Hz.)
3) A rsonance circuit in which the horseshoe magnet coils are connected to the diode capacitor so that frequency regulation is assured. The diode capacitor is then connected to the Leyden jar, transmitter unit.

The major physical principles involved in this outstanding composite unit are:

1) Electrostatic conversion using twin discs for positive output from one, and negative output from the other.
2) The evidence of the *"Searl Effect"* from the use of multiple, identical steel segments inducing an E.M.F. electro-magnets at the discs periphery (rim).
3) The *Ecklin principle* is also in evidence, since the steel segments pass by permanent horseshow magnets, as in Ecklin's S.A.G. units.
4) The *Poggendorff* self-rotating electrostatic motor principle, as described above.
5) The crystal capacitance function of the crystal diode module. The full operation of this unique component, with its hollow cylindrical permanent magnet, is a composite component with the dual functions, as described above.

The Swiss M-L Converter, — "A Masterpiece of Craftsmanship and Electronic Engineering".

Members of the G.A.G.F.E. have inspected this Swiss system on five different occasions from 1984 to the present. There are two small units and this presently described larger unit located in a commune near Bern, Switzerland. This machine and the two smaller units have been running on and off since 1982.

The larger machine produces 3 to 4 KW at 230 volts D.C., and apparently extracts energy from the gravity stressing field, and there is no primary propulsion of any kind.

This type of gravity energy field converter confirms perfectly the Bearden and Nieper model of the tachyon field. This is especially true for the considering of charge and mass of the electron to be separate. The converter runs continuously by itself, with only rotating wear parts being the two ball bearings at the center of the two discs.

The M-L Converter is functionally constructed, completely symmetrical with the two discs made of acrylic plastic, a light metal lattice, insulated copper wires, a secret crystal-diode rectifier, and gold-plated electrical connections. Everything is hand-made with the finest craftmanship, with an elegant beauty. The operating principle has been known for a long time, and these machines have been developed over a twenty year time span.

In electrostatic generators, the air molecules between the two acrylic discs which closely counterrotate, side by side, become electrically activated by friction. This causes the discs to be continually charged, until a flashover equalizes them. To limit the electrical voltage to a desired amount, the positive charged particles on one of the counterrotating discs and the negative charged particles on the other disc are each extracted by means of separately adjustable lattice-electrodes, and are fed into a Leyden jar which collects the energy. The speed of the discs, on which a fan-like structure of 50 lattice electrodes are etched out, is 60 rpm. (It is obvious that this discrete ratio of lattice/segments and speed will produce a 50 Hertz, pulsed D.C. output.) This speed is synchronized by magnetic impulses.

The unit is hand started by revolving the two discs in opposite directions, until the converter was charged up to such a degree that it synchronized itself and continued to rotate smoothly and noiselessly, without any input source of power. A centrally mounted disc of about 4 inches in diameter was glimmering in all the colors of the rainbow. After only a few seconds the Leyden jars were ready for operation, so that 300 volts D.C., with a current of 10 amperes could be extracted at the terminals, and this could be done continuously for hours, or for years, without any wear!

To demonstrate the power available, connections were made to both, alternately, a high power incandescent lamp or a heating element, each of which was rated at 380 volt service. The brilliant light from the lamp was blinding, and completely illuminated the hall to the furthest corner. The heating element became so hot, after a few seconds, that it could not be touched.

This experience was certainly a look into the future for all of us, and the start of a new era! It became evident for everybody who saw this converter functioning, that the teachings of orthodox science must undergo a complete revision in order to be taken seriously.

(The fundamental law of physics, according to Robert Mayer of 1842, is "The sum of all energy forms is constant.") Today there are already dozens of known violations of the orthodox energy laws.

This project work represents international science at work, in it finest form, which will become the wave of the future!!

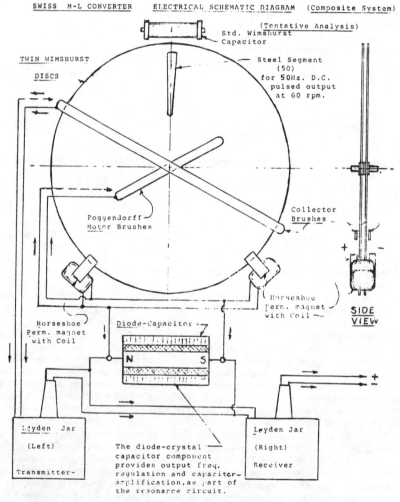

SWISS M-L CONVERTER   ELECTRICAL SCHEMATIC DIAGRAM   (Composite System)

(Tentative Analysis)
Std. Wimshurst
Capacitor

TWIN WIMSHURST

DISCS

Steel Segment
(50)
for 50Hz. D.C.
pulsed output
at 60 rpm.

Collector
Brushes

Poggendorff
Motor Brushes

Horseshoe
Perm. magnet
with Coil

SIDE
VIEW

Horseshoe
Perm. magnet
with Coil

Diode-Capacitor

N   S

+
—

Leyden Jar

(Left)

Transmitter-

The diode-crystal
capacitor component
provides output freq.
regulation and capacitor-
amplification,as part of
the resonance circuit.

Leyden Jar

(Right)

Receiver

Note: The two Leyden Jars are also a part of the resonant circuit, since one is a transmitter (Sir Oliver Lodge's experiment) and the other is a receiver and function at the same resonant frequency.

11
Unified Particle Theory

Unified Particle Theory

(Editor's note: Inventor Joe Newman has caused a major ruckus with his "magnetic particle" machine, and many people have read the inventor's theories as presented by Sam Taliaferro in **MAGNETS** *(May 1986) as well as in his self-published book. Essentially, Joe Newman theorizes that magnetic particles are gyroscopic in nature. Now comes John Griggs of Prineville, Oregon with an interesting version of the Unified Particle Theory (UPT) and we are delighted to present it to you in full. Mr. Griggs first developed his theory in 1954, and has been allowing it to grow and improve over time. We share a portion of his letter to this editor as a preface to his updated, 1985 version.)*

This paper was condensed nine years ago from one small, though important, aspect of the unified particle theory which I began an entanglement with (in September or October of 1954) due, first, to my doubts on the explanation offered by special relativity on the constancy of the velocity of light for all observers (regardless of states of motions, etc.); and, secondly, to my further skepticism on the gravitational explanation of general relativity, i.e., on the so-called curvature of space-time. How much better an attenuation shadow works! For thirty years I was laughed to scorn.

But now — during the last year and a half — most particle physicists are screaming that, at least, general relativity will have to be drastically revised if their new "superstring theories" are to work. Friends and acquaintances are now pointing out that I have had a "superstring theory" for thirty years. Even so, I have never called the UPT a "superstring theory." Nevertheless "superstringers" have copied many aspects of my theory, including non-pointlike properties of elementary particles. They now say that charge is smeared out, while with our basics this has always been a requirement. And behold! They now say that "curvature of space-time" must be drastically changed. If one could only cry! And they have even copied the torus for the internal constituents of nucleii — but not of the electron, they suppose. I'll be so bold as to say it again: These internal constituents **are** electrons (±). It is no fault of my own that they are groping in the dark yet, for I sent extracts of the UPT to the theoretical physics departments of all major universities in the United States twenty-five years ago (I have some of these returned, unopened). And at many other times I gave excerpts of the UPT to renowned physicists and astronomers in academia.

Many places in the enclosed paper I use the terms "basics of space," "basic photon flux," "magnetic flux," and "flux wind," etc. They are all the same. On page two I call these "photons." This is also true, but a further explanation should be given: These basics are photons which have split up or which have never joined to an oppositely

spinning partner.

Now that I have said this, I must say further that the electromagnetic radiation which we can sense or measure: light, whistlers, gamma radiation, etc., is made up of nutating, **oppositely spinning pairs** of our basics of space which sense and affect the basic flux. This spinning double basic photon gives a perfect picture of Maxwell's sine-waved orthogonal electromagnetic components:

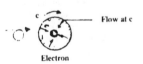

The spinning photons trace out "waves." Also see FN 10. Moreover, this double spinning photon concept explains refraction, defraction (without "waves"), polarization without ad hoc assumptions (e.g., here, "spheroidal waves" — as if there could be any such thing), and all the other known properties of electromagnetic radiation fit beautifully. This is explained in the unified particle theory, from which this paper is extracted.

But it is not proper that I should bring in all the ramifications of the statements made or ideas touched upon, or this explanatory note must be much bigger than the paper. The paper is on "elementary" particles.

A Two Component Particle Hypothesis
(Taken from the *Unified Particle Theory*)

If particles are assumed to be composite structures whose components were at one time photons one can build up a two component particle hypothesis which not only fits the phenomena of charge, spin, etc., including a new conserved quantity in all particle reactions, but also he can, by assigning a mass (when bound) to each of these two components, have a pretty near fit with the masses which are known from experiment. Others are predicted. This hypothesis is much simpler than the quark ones. And its predictions are more exact.

I make these assumptions about a photon: (1) It is cylindrical and perhaps hollow (or, depending on spin diameter, disc like or annular) in shape. (2) It spins always at c while moving linearly at c, and these two movements are retained once the photon becomes part of a particle (any point on surface traces a helix of 45°, so the trace speed is 1.4142c).

To become a particle of mass this photon must curl upon itself forming a torus. There are two generic types: closed and open. The effectiveness of the former spin (the spin 1 of the photon) is now "cancelled", though the spin is retained; the tori of charged leptons are closed:

Photon

Flow at c

Electron

The four arrows symbolize the annuled spin of the former photon; the single outside arrow, the new particle spin. The spin shown by the 4 arrows is critically necessary for energy-to-mass formation; I call this movement the "flow" of the torus. If a photon describing left hand helicity curls and forms into a particle it remains forever distinct from a torus which is formed from a photon of right hand helicity because of the *two* movements (one cannot be changed to the other). These are truly elementary. One is an e⁻; the other an e⁺. Either might now, in translatory motion, show left or right helicity (have spins parallel or antiparallel); however, one helicity would be preferred, as will be shown directly.

Charge is assumed — in this scheme — not to inhere, *per se*, in the torus, but rather to be a manifestation of the electro-magnetic radiation of space[1] (flowing through the hole of the doughnut.

I assume that space is permeated with photons of various "wavelengths". These ordinarily do not mutually interact while traversing space together any more than, say, two searchlight beams interact or, a microwave beam and photons from the sun interact. These photons collectively I call the basics of space (since they are the basic stuff from which not only matter, but charge, as well, is made).[2]

Although the photon may possibily be hollow, have a cylindrical shell structure, as a charged particle I presume that, in any case, it squeezes down to a structure filled (with the "fire fluid" of the photon).[3] Electrons should be very small (and no photon less energetic or shorter than P times its diameter can ever become an electron).

The torus contributes a mass of about 68.6 meV when *bound* in nucleii. When free, 0.5 MeV, the e+.

There has to be a second particle constructed from photons with no charge capabilities, or mass while free. It must come in two versions. It is, when free, the V_e and V_r; also, of course, the two antineutrinos. I make the assumption that perhaps the neutrino remains a hollow torus. However, unlike the electron, the tori are parted on one side. The neutrinos form spirals of permanent helicity. In sketch note that one must here consider helicity a permanent physical parameter even as one must consider a screw as having unchanging helicity; this is

aside from the usual vector definition, of "parallel" or "antiparallel" spin.

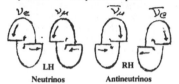

LH   RH
Neutrinos   Antineutrinos

One can have four distinct particles, two of left helicity, and two of right, fitting the requirements of V_e, V_r, V_e, V_r. This quadrality obtains because, with the permanent helicity, we have also the locked in photon spin, analagous to flow in the electron's torus. We have no V_r.

Further, since some force must hold the helices unchanging I assume this is the disturbance in the basics (an electro-magnetic 'field') caused by the two spins, and the two (hollow) faces. The V is electromagnetically neutral overall for the reason that the magnetic polarity and strength of one face (at

where the tori are split) would just oppose that of the other, neutralizing the V as a whole.

This uncharged, apparently massless (when free) lepton contributes a mass of 18.53 MeV when bound inside the composite particle.

Using multiples of the two energies and adding them together we come near the experimentally determined values.[4] Stable and metastable particles.

All 1/2 integer spin particles have odd numbers of components when summed; even integer spin ones have even numbers of components.

Note that the above scheme works reasonably well in calculating the masses of composite particles irregardless of whether they are leptons, mesons, or baryons.

The masses derived are not exact; there is an error of about 1 MeV out of 500 MeV in some. Binding energies are not uniform; and are unexplained. It is noteworthy that no other scheme comes nearly so close.

Particle	Spin	Mass MeV	n(68.6)	+	n_2(18.53)	Calculated Mass MeV
$\mu^\pm$	1/2	105.66	68.6	+	2(18.53)	105.66
π^o	0	134.96	2(68.6)	+	0	137.2
K^o	0	497.67	4(68.6)	+	12(18.53)	496.96
η^o	0	548.8	8(68.6)	+	0	548.8
$p^\pm$	1/2	938.279	11(68.6)	+	10(18.53)	939.9
Λ^o	1/2	1115.6	16(68.6)	+	18.53	1116.9
Σ^o	1/2	1192.46	16(68.6)	+	5(18.53)	1190.25
Ξ^o	1/2	1314.9	17(68.6)	+	8(18.53)	1314.44
Ω^-	3/2	1672.2	19(68.6)	+	20(18.53)	1674

If all particles are composed of these two types of tori then there should be some conserved quantity — akin to the tori — common to all composites (other than charge).[7] Call this quantity 2-toriness (2- -ness); then:

2- -ness) = Q - A, where Q is charge and A the baryon no.

Add algebraically. In *evvery* reaction 2- -ness is conserved. The e+ and its Ve are *together* given a 2- -ness of +; the e-Ve, -1, except in inverse beta-decay, where the electron alone has 2- -ness of -1 (+1).

Examples:

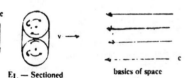

Now, above I said one helicity of the e-, e+ would be preferred. If we section one of these tori and make flow lines

E_L — Sectioned   basics of space

in the fire fluid around a *center* we may see why. We assume that the tori can have a maximum velocity relative to the basics of space of 2c (in the same sense that accelerated electrons encounter the $2°7$ blackbody radiation at near 2c). But now we must differ, for I assume that not only is this important but that it is all that is important to inertia and mass: As the electron's velocity increases into the basic flux the fluid stuff near outside of torus must meet the basics at an increasingly higher rate than does the fluid near the hole. I am assuming a torus which flies forward face on. So the center of the flow must move ever inward to keep the fluid flowing at c. If enough energy were available (an infinite amount?) this center

would reach the "skin" (at which time the torus would lose all mass and likely reverse directions). I assume mass is this movement about the center; the assymetry of the flow as seen in section is a measure of the mass. If the flow center is moved, say 90% of the distance to inside edge, this represents a certain inertia of mass, as when we throw a ball (fast ball!), and until an opposite "force" acts upon the ball — and so changes the center of movement — it will continue on.[5] The mass of, say, a proton at "rest" on the earth is about 938 MeV due to the repression of the flow centers of the three component particles within the proton. Inertia and mass are the same.

Accelerated mass is added on, therefore, without permanently changing the particle.

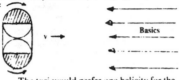

Basics

The tori would prefer one helicity for the reason that, in sectional view the hatched, the outer portion contains a greater volumn. And least action would prefer that the greater and slower flow impact the basics; the lesser flow makes up for this in speed up to near $2c^6$.

A final question should be this: do large multiples of these components exist inside the more massive elementary particles as separate entities?

It appears unnecessary in all cases since the repression of the center flow in the *closed* tori — by accelerating into the basic flux — and the increase in D, decrease in d, of the *broken* tori is the equivalent of having mass added on, where D and d are the two diameters of a torus as used conventionally. By this the scattering cross section of neutrino interactions should increase as energy goes up, that of the electron, decrease.

I have said above that one could construct "elementary" particles with only "two" torus-like components, of specified internal spins and flows.

Therein I made statements to effect that one could see the experimentally-confirmed properties of the particles by picturing these as made of the tori components. I mentioned a few.[7],[8],[9]

Now I should like to show that parity violation was no violation at all, but a misinterpretation, a confused view. The confusion was partially, at least, caused by one's having pictured the particle's spin as either "pointing forward," right hand, or, "backward," left hand. Arrows pointing opposite to the direction of the particle's linear movement are not only misleading but are one

dimensional affairs; whereas, we need arrows showing *movement* in *three* dimensions (arrows pointing with the direction tell little more). It is small wonder that in viewing the beta decay of Co^{60} in terms of the conventional arrows we should ascribe to the mirror an impossible event.[10],[11] Convenient mathematical fictions can cause trouble. In terms of the traced movements and helices of our tori there are not only no "impossible" mirrored events but the events as mirrored are the requisite antiparticles. (When we have no translatory movement we must yet picture spin and flow — but not helicity.)

Let us use mirrors.

Sectioned Components

Mirror   Images

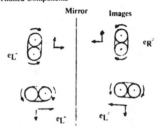

Consider the unbroken tori, which I assume are the electrons: If a torus of one flow and spin (with a preferred direction of translation) constitutes an e- then its mirror view with all the opposed movement, of opposite helicity, is the e+. The mirror view of the e- is not still an e- ("since we can have left and right electrons," as it is usually said), for we must consider flow also. The torus can show left and right helicity, *but* to be an e- with *left* helicity we must have outside to forward flow,

flow   spin
e_L
                Trans-

                Trans-
e_R
          spin

and an e_R^- must show opposite, i.e., inside to forward flow.

The mirror views of these two are the two helices of the e+. The flow center configuration should prohibit the e- from remaining right handed. If it is started out from a composite as right handed it should flip immediately, reacting against the basics of space, and conversely, the e+.

Images of the above
e_L^-, e_R^- become $e_R^=$, $e_L^=$

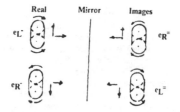

Real　　Mirror　　Images

eL^-　　　　　　　　$eR^=$

eR^-　　　　　　　　$eL^=$

If we show the more complex proton, its mirror-image will likewise be its anti-particle.

Because of the flow — and impingement of basics of space against this flow (maximum relative of 2c) — there is a preferred helicity of the electron. If, therefore the electron is treated with a magnetic field[10],[11] the electrons will assume orientations which will preferentially, at unstable particle decay, fly off in predetermined directions. consider a mirror view of π^- and μ^- decays:

netic basic flux hitting the components of Co^{60} will tend to cause the outside flow of the unbroken tori (within the Co^{60}) to face into a helically aligned flux "wind" (electromagnetic field) — exactly as when the tori are accelerated. This aligning of the tori (since they do not flip except under special conditions of breakup of composite) is that which aligns the whole:[10]

Magnetically aligned basic flux

eL^-　　Spin alignment symbolized

When these electrons are "shot" out they are shot out so that the outside flow heads into the flux. This repression of the flow center *is* inertia; when any weak spot (or hole) in the impinging flux passes by, the component,

claim that the principle of time reversal is anything but invalid: we can start at the macro level, using the illustration of a swimmer moving backward in time (a favorite of the time-reversal-*in principle*-crowd). As the swimmer moves forward in time we see, by close scrutiny, that there is a reaction between the water and the swimmer's feet and hands. But in the backwards view — and we must see all — the water molecules rush up ag15t the swimmer's hands and feet, stop dead, causing no reaction; instead the swimmer moves back along the path from which the water came. Action-reaction is violated, Newton's third law of motion.

Let's go to the very small: we have a π^+ moving through a magnetic field, the magnet has poles, and marked N and S; the magnetic field is of such a direction, say, that from our vantage point the π^+ curves to the left (a π^- will curve to the right). If we run a film of the event backwards the π^+ will be seen to curve right, an impossible in our magnetic field. Ah! but let the camera reverse left for right. Now our π^+ curves left again. But perversity has us, for when our camera changed left for right it also changed all the internal spins and flows of the π^+. And this changed particle is a π^-. A π^- should curve to the right, not left. But wait! did our camera not also interchange the north and south poles by reversing the spin of spin-aligned basics traversing the space between poles (despite the now phony fact that the south is imprinted N, the north, S), reversing our field? Indeed. Now we have the π^+ curving left, well enough, but since our magnetic field is reversed, it should now curve right, not left. Therefore time reversal is impossible. When we have allowed ourselves the abilties to see the spin and sub-structures of all components of a particle we see that we have changed a particle into its antiparticle, we now have a π^- curving correctly in a reversed field, but we have not run the π^+ backward in time. All such experiments with micro (as with macro) constituents fail if we are allowed analytical instruments, e.g., magnets in determining all parameters. Only if we disallow instruments showing charge (the curve), ionization, spin, momentum, etc., may we say: The most elementary particles can travel backwards in time — unfortunately we are hiding behind our ignorance. For when we allow total analysis, showing the known properties of particles, no example of time reversal is seen. Indeed since one can show that time reversal is impossible in the four simple basic particles (by our hypothesis), the electrons and neutrinos, due to its violating one or more of the known laws of nature and

π^- spin 0　　Mirror　　　　π^+ spin 0

V_{MR}　eL^-　V_{er}　$V\mu L$　　$V\mu R$　VeL　eR^+　$V\mu L$.

At breakup　　　μ^- spin 1/2　　　μ^+ spin 1/2

Some flip at breakup　　e^-　Ve　$V\mu$　　　$V\mu$　Ve　e^+

L　R　L　　R　L　R　　　L

All internal movements are reversed by the mirror; therefore the images are the antiparticles. Since π^- has spin 0 and the μ^- has spin 1/2 we must assume that when the π^- breaks up into a μ^- and an $V\mu$ the e^- flips over from a state of left spin to one of right spin; this allows the $V\mu$ and μ^- to fly apart with the right hand helicity. The flip takes time, slowing the reaction. Nevertheless, except to preserve angular momentum when the need arises, the flow pattern of the e^- should make it left handed.

Finally, if we align a composite particle by using an electro-magnetic "field" (our helically aligned space basics) as Mrs. Wu and associates did with the Co^{60} nucleii[12] we thereby align the composite in a preferential way since the *individual* components are all that offers a resistance to our spin aligned basics, indeed, all that there is to any particle. All should pretty much agree with that. Now, the important thing is this: the mag-

the e^-, must shoot out — the steam is up, and throttle is open. The chuck is removed.[13]

What of charge, of the so-called TCP theorem? We have treated it above, already; for when we treat spin, flow, and helicity (through the basics) we have simultaneously treated charge. Charge is the twist given the basics by spin and flow. It is then small wonder that "weak interactions always obey charge-conjugation invariance and parity invariance taken together". They are inseparable. The parity (true parity) determines the charge.[14]

And we treat not just charge-conjugation invariance and space-inversion invariance, or parity, but, at once, time-reversal invariance. For when we allow that any parity experiment must see all internal constituents and their several movements — and without this allowance we are hobbled, have blinders on — we at once see that no experiment, including "thought" experiments, can since (in our view, above) all matter is composed of these can we not conclude that time reversal is disallowed in all more complex structures by existing laws of nature?[15]

FOOTNOTES

1. Besides the 2°.7 blackbody of Penzias and Wilson and the other known electro-magnetic radiation of space I assume there to be much other. More than 90% of the substance of the world is unaccounted for; it is needed to account for the formation of the gravitational systems. If this missing mass is in the form of photons we should have enough.

2. The elementary particle aspect of the UPT with which this paper is concerned was first conceived in about 1957, excerpts of which were printed in pamphlet form (on three occasions in the early 1960's). New data was added as it became available. I call the overall scheme the unified particle theory, and intend publishing the concept in book form. As far as I can determine the French-Swiss LeSage was the first to propose a particle shadowing effect gravity.

3. Though a filled in cylinder is not apparently a prequisite here, it might give us a clue as to why the Bohr energy orbits contain the numbers: 1, 4, 9 ... 36 (Balmer and other series). If R, r are the lii (in usual sense) of the torus; then in maintaining a constant volumn as r assumes values 1, 1/2, 1/3 ...; R assumes the values 1, 4, 9...

4. I apologize if due credit has not been given anywhere. I have little access to the literature.

5. I assume that both the broken and unbroken tori have the ability to replicate themselves, given enough energy through acceleration. The new material is from the basics of space. But no torus can replicate unless it has basics on both sides (and acceleration sees to this) thereby producing pairs of components of opposite spins and flows — all the intrinsic properties there are (all other properties are seen as caused by these spins, and flows, and, their interactions with basics).

7. The electron charge remains constant due to constant volumn, spin, and flow of the torus.

8. Although the numbers and kinds of particles and resonances are, by this hypothesis, restricted in the lower mass range there are nearly unlimited possibilities at higher and higher masses, given sufficient energy of acceleration and some degree of stability.

9. Since I have made the assumption that nothing exists except the basic photons of space the two tori, broken and unbroken ones, then all "forces" of nature, gravity, weak, electromagnetic, and strong must ensue from these. It is not difficult to envision that tori in proper juxtaposition would be difficult to break apart because of helicity of basics (charge) flowing through the holes of tori whereas at some distance we might

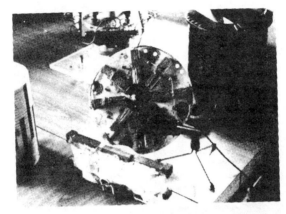

Over Unity Motors??

have repulsion. Further, certain positionings can be seen as readily breaking apart, i.e., being pushed apart; others only slowly, only after having flipped over — recall, if you will, that we have a preference of helicity for the closed tori; and a permanent helicity for the broken ones. And gravity can be seen as a push shadowing gravity, much like the one LeSage envisioned. The total interactions (weakening) of the basic flux is the important criterion here. We need no "field" - - the weakened flux density *is* the field.

10. In our view an electric "field" is the alignment of spins of basics *caused* by the flow of electrons through the basic stuff of space; a magnetic field is the spin allignment of basics caused by the flow of the basic stuff through electrons (which are alligned and usually "stationary"). In both cases the basics go through the hole of the torus, i.e., when the tori move past the basics we have the electrical, and when the basics move past the tori we have the magnetic one.

11. In the famous parity overthrow experiment suggested by Lee and Yang.

13. Radioactivity, its randomness, and tunnelling are naturally explainable by slight inhomogeneities in the basic flux. One should not stop here. Both the high energy production and superluminal velocities of QSO's can be seen as caused by matter (galaxies) having reached near the edge of a bounded, steady — more or less — ball of basic flux. The nearer one comes to the edge the more particles that become unstable. Out of the flux *all* are unstable. Total internal reflection will keep the size of the world constant. At edge of basic flux, matter all across a galaxie, without regard to extent, could simultaneously break up to radiation — hence the "superluminal" velocities.

14. If this view (from the unified particle theory) is correct then the v, while possibly massless, should not of necessity have a linear velocity of c. By producing short bursts of neutrinos one should be able to time the flight of the bunches. A test. An alternative to this view of the v, and one which I once considered, was this: It is a mold pattern molded in the basic flux by reactions against that flux by the tori — a pattern molded by kick-back at time of break up of composites, so it would be a dimple in the flux of specified energy and spin angular momentum.

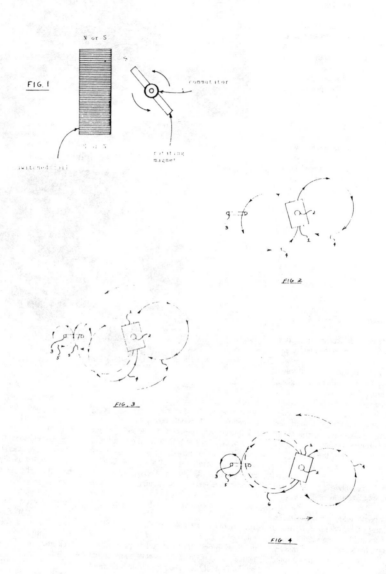

FIG. 1

N or S

commutator

rotating magnet

switched coil

FIG 2

FIG, 3

FIG 4

INTERACTION OF COIL STRUCTURE AND ROTATING, CYLINDRICAL, PERMANENT MAGNET

WIRE FROM
BATTERY TERMINAL (−)

The expansion and
collapse of the
coil's magnetic
field can also be
considered inde-
pendently of the
cylindrical perm-
anent magnet. In
addition, the
cylindrical perma
nent magnet could
be placed inside
the coil structure
If this is done, then
to avoid a "cancellation
effect" the width of the
opening on the coil structure should **exceed** the
end width of the cylindrical
permanent magnet.

WIRE TO
BATTERY TERMINAL (+)

MAGNIFIED VIEW OF COPPER WIRES

TOP
(looking down from above)

Input current direction

spiral-helix path of gyroscopic particles (only a few depicted)
which generate the magnetic field around the coil

---

(1) "collision" occurs between the **upper periphery** of the gyroscopic particle and the adjacent copper wire

some gyroscopic particles
"miss" the wire and
physically expand
beyond the wire

gyroscopic particle
now travels in (−) direction,

expand

expand

input direction

input (catalytic)
current direction to
initially align the atoms
within the copper coil

[When this happens, the
gyroscopic particles physically
expand away from the atoms of the wire
from which the gyroscopic particles emanate.]

MOMENT OF FIRING SEGMENT ON THE COMMUTATOR
(expanding magnetic field)

---

(2) now the gyroscopic particle "collapses" to return to the atom from
which it emanated — but now the **lower periphery** of the gyroscopic
particle first strikes the copper wire causing the gyroscopic particle
to now travel in the (+) direction

collapse

copper coil composed
of copper atoms with trillions
of gyroscopic particles

MOMENT OF BLANK SEGMENT ON THE COMMUTATOR
(collapsing magnetic field)

---

(3) The process below is fully repeated with step (1) physically thrusting the expanding
magnetic field even further beyond the coil structure [This expansion and collapse
of the coil's magnetic field mechanically interacts with the magnetic field
(gyroscopic particles) of the cylindrical magnet to create a "push-pull" effect which
causes the cylindrical magnet to rotate and also add some of its gyroscopic particles
in interaction with the copper coil thus adding further to the external output energy.]

expand

gyroscopic particles traveling in (+) direction
after reaction (2) once again align
the copper atoms causing the
magnetic field to
expand again

MOMENT OF SHORT OUT SEGMENT ON THE COMMUTATOR
(cancelling additional current from the battery)

---

not to scale — qualitative depictions only

Telephone No.: 5514684
Telegrams : "ATOMPOWER"
Telex   : 011-2510

भारत सरकार
GOVERNMENT OF INDIA
परमाणु ऊर्जा विभाग
DEPARTMENT OF ATOMIC ENERGY
न्यूक्लियर विद्युत बोर्ड
NUCLEAR POWER BOARD
500 MWe PHWR

एस-७१, साउथ साईट,
S-71, South Site
ट्राम्बे, बम्बई-४०० ०८५,
Trombay, BOMBAY-400 085.

P. Tewari
Head, Quality Assurance

D.O. No: 005/00000/86/B/   November 11, 1986

Dear Mr. Soule,

    I have received your letter of 14th Oct, 1986
regarding invention of a energy machine by Joseph
Newman. I very much believe that such an invention
is technically possible. That the patent office
has not granted Mr. Joseph Newman, patent for his
invention despite his 7-year struggle in this regard
as stated in your letter, is indeed a matter of
regret. I am enclosing here a copy of my latest
works 'Beyond Matter' which lays foundation for
generation of energy and also matter and from
absolute vacuum. Any one who doubts the theoretical
validity for a machine of the kind discovered by
Joseph Newman, and was learned by me through the
relevant literature sent by you to me can certainly
go through my works and write to me for detailed
discussions.

    I shall certainly extend my help in whatever
way you deem fit. Please do write to me a
specfic plan in assisting Mr Joseph Newman if you
have any.

    With best wishes,

                                      Yours sincerely

Encl: as above
                                      (P. Tewari) 11/11/86

Mr Evan R. Soule, Jr.
1135 Jackson Avenue, Suite 305,
New Orleans, Louisiana 70130
(504) 524 - 3063.

JOSEPH NEWMAN'S TECHNOLOGY - Multi Field Coil Concept

Joseph Newman has something more to teach us with Figure 6, below, of his South African Patent!!

Simply stated, he is showing us that we can use our input E.M.F. and apply it too!!

To clarify this above statement, Newman's S. A. patent art discloses that the initial E.M.F. can be used to run the motor portion of his unit (300) in the drawing, below, while the electrical field effect increases the magnetic field of the motor coils, (303), as was previously discussed and illustrated.

The motor field coils (305) in turn, can be inductively coupled to one, two, or as an optimum of three induction coils, which then become E.M.F. generators, as (306) which will operate with no back-EMF involved.

The multiple induction field coils (306) are illustrated below in schematic form.

This is a very significant and momentus development in electrical engineering and specifically in electrical motor/generator design, which has not as yet had its full impact in the o/u/o field.

A minimum of threefold electrical output over unity ratio can easily be expected with the maximum ratio yet to be determined. It is projected that the maximum output to input ratio could be as high as twenty-five to one, by the use of optimum field coil design, that is the use of at least three secondary field coils, as indicated, and the idealizing of all the functional components of the unit.

From Bruney Research:- Any increased drain on the secondary coil windings will cause an increased current drain in the primary windings. However, the greater the current drain in the primary coils, the stronger the interaction between the primary coil and the permanent magnet (rotor). The result is that the increased current drain increases the output of the machine. The maximum output is, in turn, limited by the field strength of the permanent magnets.

In summary, the addition of the multiple secondary coils should effectively increase both the output and efficiency of the machine.

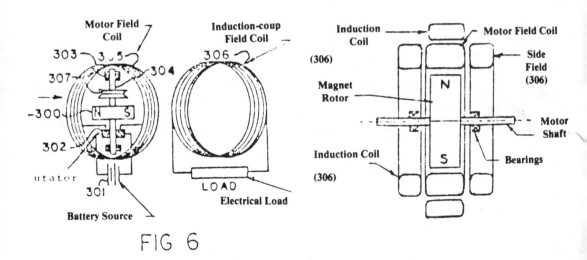

FIG 6

Statement of Roger Hastings, Ph.D.
Before the Subcommittee on
Energy, Nuclear Proliferation, and Government Processes
Dated July 30, 1986

THE NEWMAN CONTROVERSY

My name is Roger Hastings. Thank you for the opportunity to address this Committee. Before I discuss Dr. Newman's device, let me tell you something about my background, how long I've known Joe Newman, and then I'll make my observations about Newman's device, and repeat my criticisms of the recent N.S.B. tests of Newman's device.

I. BACKGROUND

My education, through the Ph.D., is in physics. I have served as a Professor of physics for four years, and for the past five years I have worked as a physicist for the Sperry Corporation in St. Paul, Minnesota. My current title is Senior Staff Scientist, and I am the manager of Sperry's Superconductive Electronics Technology Center. I have known Mr. Joseph Newman for five years. During this time I have tested most of the many prototype motors which he has constructed, and I have witnessed testing by other technical people. I have become familiar with Mr. Newman's theories and attitudes. I represent myself in this matter, and have never represented Sperry Corp. regarding Mr. Newman or his machine.

II. NEWMAN'S DEVICE

Newman's motors all consist of a very powerful permanent magnet which rotates or reciprocates within or near a coil consisting of a very large number of turns of copper wire. The coil is energized by a battery pack, and the magnetic field produced by the coil provides the torque or force required to rotate or reciprocate the permanent magnet. A mechanical commutator reverses the direction of current flow through the coil each half cycle, and in some models also chops the current input between current reversals. Technically, the motor may be described as a two pole, single phase, perma-

nent magnet armature, d.c. motor. The difference between Newman's design and the prior art is one of scale: very large magnet and very large coil. Newman's large motors contain conventional ceramic magnets weighing up to 700 pounds. His smaller motors use powerful rare earth mgnets. The coils typically are wound with more than 100,000 turns of copper wire. Since the coil resistance is therefore high, the machines operate on battery voltage which is sufficiently high (hundreds to thousands of volts).

The torque applied to the magnet in these motors is proportional to the product of the strength of the magnet, the number of turns of copper wire, and the current flowing through the wire. In Newman's machines extremely large torques can be developed with very small current inputs. If we scale up Newman's motor, it is theoretically possible to obtain infinite torques with infinitesimal current flow (and not violate any laws of physics). However, according to conventional thought, as soon as this magnet began to rotate, doing work against some load applied to its shaft, the back emf (electromotive force) produced by the rotating magnet would produce a back current which nearly cancels the input current, and the torque would be reduced to nearly zero. The magnet could not rotate, or would rotate extremely slow with the shaft power output less than the battery input.

Consider what has happened to conventional thought in the past when people have experimented with the limits of very high speeds (relativity), very small dimensions (quantum mechanics), very low temperatures (superconductivity and superfluidity). Newman's motors probe the limits of very large torque with very small current input. And they do rotate at relatively high rates. For example, witness Newman's latest prototype (on

demonstration following this hearing today in an auditorium in this building), which runs on 0.0008 amps at 3000 volts and turns a 16-inch fan blade at more than 500 r.p.m. How much torque can this motor produce? Try to stop the motor by holding the two-inch daimeter shaft. This will not be possible for a normal human, although the motor will never draw more than 0.003 amps or nine watts. This motor is a scale model of a motor which Newman intends to build to power an automobile.

Newman's motors are unconventional in other ways. One notices the flourescent tubes which are placed across the motor coil. These tubes are lit by the coil's collapsing magnetic field occuring when the battery voltage is switched. They are used to protect the mechanical switch from damage due to arcing. The additional power produced in these tubes (and flowing through the system) occurs at very high frequencies, primarily in the range of ten to twenty million cycles per second. This r.f. (ratio frequency) current has been accurately measured, and exceeds the battery input current by factor of five to ten in the various motors. One of Newman's motors was monitored with a computerized high-speed data sampling system, with the following results:

(1) The r.f. appears in bursts, with the repeat time between approximately equal to the length of the motor winding divided by the speed of light in copper. The r.f. bursts showed little attenuation during their travel through the coil, maintaining their shape and amplitude.

(2) The r.f. current and voltage were in phase, representing the real power.

(3) The r.f. current and voltage wave-forms were offset from ground, indicating a net d.c. component.

net d.c. component.

(4) The net r.f. power at the battery pack represented a negative power which exceeded the d.c. input power from the batteries.

The last statement may explain why Newman has been able to demonstrate the charging of dry cell batteries placed in his system. Battery failure has occurred through internal shorts which develop within the batteries rather than by depletion of the energy stored within the batteries. When you witness the demonstration of Newman's latest prototype, if you attend the demonstration following his hearing, bear in mind that the batteries will last many times longer than expected for a drain of 0.0008 amps. A prominent battery company is working with Newman to develop batteries which will stand up to the r.f. power levels, and perhaps last even longer.

Newman's motor design is based on his theory of gyroscopic particles which he explains in his book "The Energy Machine of Joseph Newman." Full utilization of his machine will require a detailed mathematical representation of the phenomena based on a thorough understanding of the atomic processes at work. This will require a parallel program of experimentation using the finest resources available. Application programs have already been conceived (for example, the car motor), and will require prototyping and manufacturing efforts. Newman should be immediately awarded a patent and become recognized in the scientific community for his accomplishments to date.

III. AN EVALUATION OF N.B.S. TESTING

I have been asked whether the recent N.B.S. tests alter the opinions I've expressed before and I'm repeating here today. The recent N.B.S. tests don't alter my opinion because N.B.S. failed to test Newman's device.

I have read and evaluated the Newman motor test results reported by R. E. Hebner, G. N. Stenbakken, and D. L. Hillhouse in National Bureau of Standards Report #NBSIR 86-3405. [See "Report of Tests on Joseph Newman's Device," U.S. Dept. of Commerce, dated June 26, 1986, hereinafter referred to as "the NBS Rpt. at ____".]

A. N.B.S.'s Energy "Output" Measurements

While the reporters display fine credentials and demonstrate the use of precision equipment, they obviously did not test the Newman motor. Instead they measured the power consumed in resistors placed in parallel with the Newman motor, and called

this power the motor output. [See NBS Rpt. at 7. Fig. 4. "Schematic Drawing of Newman device and input and output power measurements circuits," reproduced with comments plainly referring to the "Resistors" as such in the accompanying chart.]

In layman's terms, this is equivalent to stating that the output of an electric motor plugged into a wall socket is given by the power used by a lightbulb in the next room which is on the same circuit. The measurement of power consumed by these parallel resistors is clearly irrelevant to the efficiency of the Newman motor.

The actual input power to the Newman Motor (battery input minus power consumed by their resistors) is referred to in the report as "internal losses." No attempt was made to measure the mechanical output of the Newman motor. Nor was any measurement made of heat generated in the motor windings.

B. The Additional Energy N.B.S. Lost From The System

It has been demonstrated by myself and others that much of the excess energy generated in the Newman machine occurs at very high frequencies (in particular between 10 and 20 MHz). It has also been demonstrated that the high frequency current will flow to the ground if given the opportunity. If Newman's machine is grounded through a high resistance, heat will be produced in the resistor which represents an additional motor output. In the N.B.S. testing, the Newman motor was connected directly to ground, thus eliminating the excess r.f. power from the system [See NBS Rpt. at 7 (Fig. 4.) "Schematic Drawing of Newman device and input and output power measurement circuits," reproduced with comments plainly referring to the "Ground" as such in the accompanying chart.] The report states that "the power flow in the device is primarily a low frequency phenomena." This result was guaranteed by the test set up. Again, the oscillographs shown on page 3 of the report show clean low frequency waveforms. All oscilloscope waveforms which I have observed on Newman motors which are properly connected, have by contrast been dominated by extremely large high frequency components.

C. Conclusion

In conclusion, the N.B.S. failed to measure the output of the Newman motor, and instead measured the output of parallel resistors. In addition, the primary r.f. energy generated by the machine was shunted to ground. Their measurements are therefore irrelevant to the actual functioning of the Newman device. These results reflect a total

lack of communication between the N.B.S. and Newman or any other expert on Newman's technology. Considering the importance of Newman's machine and its potential applications, this waste of N.B.S. resources and misrepresentation of Newman's device is an insult to those seriously interested in the machine and to those who may benefit by its future applications.

Preliminary Analysis of Newman Machines

Abstract

The essential efficiency mechanism within Newman Machines are the motions of flux lines either *perpendicular or opposite* to the rotation direction of a permanent magnet. Traditional equal and opposite CEMF (counter electromotive force) losses are circumvented, and the permanent magnet is made to perform work via inductive interaction.

General Description

FIGURE 1 shows the basic components of a Newman Machine, consisting of a stationary conducting coil, a rotating permanent magnet, and a commutator which rotates with the magnet.

About 24-28 times per 360° of magnet rotation, the commutator alternately switches drive current from a battery to the coil, then disconnects the drive current and series-connects the coil to an electrical load. The switching takes place rapidly, as a spark jumps across the commutator gaps for each switching event. The commutator also reverses the direction of the drive current to the coil every 180° rotation.

The sequence of events within the device are:

1. Energy, in the form of electric current from a battery, is supplied to the coil. As a result, one would expect:

a. Part of the input energy is invested in a magnetic field which forms around the current flowing within the windings of the coil: and

b. Part of the input energy is invested in the rotation of the magnet, as a result of the interaction between the permanent magnet and the field around the coil.

2. Electric current from the battery to the coil ceases. The coil is immediately connected to a series electrical load. One result is that:

a. Part of the input energy, stored in the coil magnet field, is delivered through the electrical load as the magnetic field collapses.

If one considers only the induction action of flux lines rotating with the permanent magnet, one would expect that:

b. The remaining part of the input energy, invested in magnet rotation, induces a current in the coil, which gives rise to an equal and opposite magnetic field around the coil that directly opposes the rotation of the magnet.

The above results, however, do not reflect perpendicular or opposite motions of the permanent magnet flux lines relative to the coil windings. These flux motions are shown schematically in FIGURES 2-4.

In FIGURE 2, a permanent magnet, 1, is free to rotate around pivot 2, under the influence of a coil of wire, 3. In these figures, a single conductor of the coil is shown for simplicity, but in practice many windings are used. The permanent magnet's lines of flux are shown in FIGURE 1 by curved arrow-lines, 4. In FIGURE 1, no field is shown around winding 3, as no current is flowing in the winding.

In FIGURE 3, the magnet is shown during the first 90° of rotation, with a drive current flowing through the winding that generates a magnetic field around the winding, as shown by arrow-lines 5.

This winding field is of the same magnetic polarity as the magnet, and causes the magnet to rotate due to mutual magnetic repulsion. This mutual repulsion also causes the magnetic flux lines of the permanent magnet to be pushed inward and rotated forward of the magnet as shown. The forward displacement exists because the energy transfer between the drive current and the rotating magnet is impeded by the moment of inertia of the magnet. The inertial mass cannot respond to instantaneous drive current changes, so not all the electrical input energy is effectively transformed and stored as kinetic energy of rotation. Contrarily, the magnetic flux lines of the magnet are displaced instantly by instanaeous changes of the same drive current, and therefore act as an energy storage means for that portion of the input energy which causes the deformation.

If the drive current to the coil suddenly ceases (as it does in Newman Machines), the magnetic flux lines of permanent magnet expand outward and rearward to their original shape, releasing the energy stored during their deformation.

The outward perpendicular expansion of the flux lines induces current in the winding as the flux lines cut across the conductors. The induced current is in a direction which magnetically opposes the advancing field of the permanent magnet; that is, the perpendicularly induced current is in the same direction as the original drive current.

In FIGURE 4, the magnet is shown during the second 90° of rotation, with a drive current as above passing through the winding.

In this quadrant, the winding field is opposite to the polarity of the permanent magnet, and causes the magnet to rotate by magnetic attraction. This mutual attraction causes the magnetic flux lines of the permanent magnet to be pulled outward and rotated forward of the magnet, as shown by arrow-lines 6. This stretching of the field again acts as an energy storage means for that portion of the input energy which causes the stretching.

If the drive current to the winding suddenly ceases (as in Newman Machines), the magnetic flux lines of the permanent magnet retract inward and rearward to their original shape, releasing the energy stored during their deformation.

The inward perpendicular retraction of the flux lines induces current in the winding as the flux lines cut across the conductors. The induced current is in a direction which magnetically attracts the receding field of the permanent magnet; that is, the perpendicularly induced current is, again, in the same direction as the original drive current.

The cycle is repeated for the third and fourth 90° quadrants of rotation, only with the drive current direction reversed.

Joe Newman's Philosophy: 'If It Can't Be Done, It Interests Me.'

PHOTO BY MARY ANN WELLS

Joe Newman

A great thinker is seldom a disputant. He answers other men's arguments by stating the truth as he sees it.

— Daniel March

Joseph Newman is an original thinker who has the ability to visualize the mechanical essence of what he evaluates.

For the past 23 years, he has made his living by inventing. He holds eight patents for inventions which include plastic-covered barbell sets, a mechanical orange picker, a bike that does "wheelies," a knife that always lands point forward, and a new type of automobile rain-deflector.

Joe's formal education ended after his junior year of college so that he could pursue his true love — inventing. He did, however, continue to use his own brains and books to teach himself physics, chemistry, astronomy, and other areas of science.

At one point, Joe began to study the experiments and writings of the famous English scientist Michael Faraday, whose ideas led to the development of the modern electrical generator. Following 15 years of independent study, Joe explained his theory in a 133-page document which described a new technical process for understanding and utilizing a source of unlimited energy.

According to previously written statements, the Newman energy generator (its formal name) works successfully because "all mass is made up of electromagnetic energy, and if the proper, mechanically-designed mechanism is built, one can change mass into pure electrical energy and/or rotational motion in a 100-percent conversion process." His three operational prototypes — one small, one medium in size, and a third weighing over 5,000 pounds with a 600-pound rotating magnet — have already been constructed. He built the three machines several years ago in his workshop, and he repeatedly emphasizes that he developed his conceptual theory over a period of 15 years before building the first physical prototype.

Joe Newman says that his invention is far more efficient than conventional nuclear energy and doesn't release any harmful radiation. He also says that his invention will replace all other forms of energy. It will be utilized in the home, by industry, and to produce commercial electrical energy at a small fraction of today's cost.

More than 30 competent physicists, electrical engineers, and technical individuals have signed Affidavits which state that Joseph Newman's invention does what he says it does: the external energy output exceeds the external, electrical energy input. This additional energy output is the result of the internal, magnetic energy within Joe's invention being converted to electrical energy. [This process is described in this book.]

Those who have signed the Affidavits include Milton Everett (biomass energy specialist from the Mississippi Department of Energy and Transportation), Mike Meatyard (electrical engineer for the Alabama Corps of Engineers), Eike Mueller (European Space Agency mission coordinator with NASA), and Dr. Roger Hastings (a principal physicist for Sperry-Univac, Inc., St. Paul, Minnesota).

In his endorsement, Dr. Roger Hastings writes: "To date we have been pouring huge funds into nuclear fusion in search of this dream. It appears that Mr. Newman has found the solution on a scale which will allow immediate and economically reliable development."

As Joe Newman says, "The finished prototype of what I teach will change the world drastically for the good of mankind, more so than any invention before this time."

For the first time in our history, this book — THE ENERGY MACHINE OF JOSEPH NEWMAN — discloses the principles and technology required to create a totally new method of energy generation via "an invention whose time has come."

12
A Selection of Patents

John Ecklin   Alexandria, Virginia   & *FLUX SWITCH ALTERNATOR*

John Ecklin's "stationary armature generator" (S.A.G.) concept has inspired and produced a variety of spin-off designs, based on the principle of circumventing back E.M.F. as established by Lenz's Law.

The 1975 Ecklin U.S. Patent No. 3,879,622, spurred interest in his concept which essentially comprised of spinning a soft iron sheild or shutter between two permanent magnets thereby interrupting the magnetic lines of force. The shields serve to reverse the magnetic within stationary central coil, and thereby circumvent the back-EMF per Lenz's Law, to produce an increased electrical output yield from the coil.

These units, known as SAG's or stationary armature generators led to the design and construction of larger and more effective types of unit and eventually the V.R.G.'s or variable reluctance generators/alternators.

The Patent Abstract to the original Stationary Armature Generator outlined as follows: "A permanent magnet motor in one embodiment utilize a spring-biased reciprocating magnetizable member positioned between two permanent magnets. Magnetic shields in the form of rotatable shutters located between each permanent magnet and the reciprocating member to alternately shield and expose the member to the magnetic field thereby producing reciprocating motion. A second embodiment utilizes a pair of reciprocating spring-biased permanent magnets with adjacent like magnetic poles separated by a magnetic shield which alternately exposes and shields the like poles from the repelling forces of their magnetic fields."

The V.R.G. generally consists of alternate A.C. and D.C. field windings, with the smaller D.C. input producing a large A.C. output which can be utilized for a variety of load devices. These units have reached an input to output ratio of approximately 3:1, with further development work continuing. Both here and abroad (Denmark) and R & D effort is moving ahead on these V.R.G. units.

United States Patent [19]

Ecklin

[54] **BIASED UNITIZED MOTOR ALTERNATOR WITH STATIONARY ARMATURE AND FIELD**

[76] Inventor: John W. Ecklin, 6143K Edsall Rd., Alexandria, Va. 22304

[21] Appl. No.: 392,102

[22] Filed: Jun. 25, 1982

[51] Int. Cl.⁴ H02P 7/66; H02K 47/04
[52] U.S. Cl. 318/140; 318/138; 318/149; 318/153; 310/113; 310/155
[58] Field of Search .... 318/140, 141, 142, 144, 318/148, 149, 151, 152, 153, 138; 310/159, 102 R, 103, 113, 158, 159, 152, 154, 156, 168, 171, 177, 179, 46, 181, 155; 322/39, 90, 100, 13

[57]   **ABSTRACT**

A unitized (single unit) motor and flux switch alternator having stationary field, armature and motor windings which provides a magnetic path for some of the motor input power to feed through and increase the alternating current (AC) generator output. A rotor formed from a material having a high magnetic permeability (solid or laminated soft steel) is controlled in speed by controlling the magnitude and timing of the pulsed direct current (DC) supplied to the motor windings which may be wound on the stationary legs or the rotor. The current flow in the motor windings can be controlled by a mechanical commutator if the motor windings are on the rotor or by a solid-state converter if the motor windings are on the legs in a manner normally associated with brushless DC motors. The DC windings of the flux switch alternator can be replaced by permanent magnets since the reversing field in the AC output windings are predominantly time stationary.

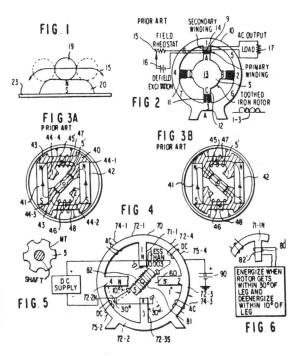

FIG. 1

FIG 2

FIG 3A
PRIOR ART

FIG 3B
PRIOR ART

FIG 4

FIG. 5

FIG 6

ENERGIZE WHEN ROTOR GETS WITHIN 30° OF LEG AND DEENERGIZE WITHIN 10° OF LEG

United States Patent [19]

Ecklin

[11] Patent Number: 4,567,407

[45] Date of Patent: Jan. 28, 1986

[54] BIASED UNITIZED MOTOR ALTERNATOR WITH STATIONARY ARMATURE AND FIELD

[76] Inventor: John W. Ecklin, 6143K Edsall Rd., Alexandria, Va. 22304

[21] Appl. No.: 392,102

[22] Filed: Jun. 25, 1982

[51] Int. Cl.⁴ H02P 7/66; H02K 47/04

[52] U.S. Cl. 318/140; 318/138; 318/149; 318/153; 310/113; 310/155

[58] Field of Search .... 318/140, 141, 142, 144, 318/148, 149, 151, 152, 153, 138; 310/159, 102 R, 103, 113, 158, 159, 152, 154, 156, 168, 171, 177, 179, 46, 181, 155; 322/39, 90, 100, 13

[56]   References Cited

U.S. PATENT DOCUMENTS

1,730,340	10/1929	Smith	322/27
2,217,499	10/1940	Smith	322/27 X
2,279,690	4/1942	Lindsey	310/46
2,505,130	4/1950	Maynard	310/155
2,520,828	8/1950	Bertschi	310/155
2,732,509	1/1956	Hammerstrom et al.	310/168
2,816,240	12/1957	Zimmer	310/181 X
3,009,092	11/1961	Carmichael	322/17 X
3,010,040	11/1961	Braun	310/181 X
3,253,170	5/1966	Philips et al.	310/168
3,346,749	10/1967	Shafranek	310/181 X
3,512,026	5/1970	Tiltins	310/168
3,518,473	6/1970	Nordebo	310/168
3,569,804	3/1971	Studer	318/138
3,577,002	5/1971	Hall	310/240 X
3,588,559	6/1971	Fono	310/168
3,594,595	6/1971	Frederic et al.	310/168

3,641,376	2/1972	Livingston	310/113
3,879,622	4/1975	Ecklin	310/80
3,953,753	4/1976	Barrett	310/168
3,967,200	6/1976	Tetsugu et al	310/113 X
4,053,801	10/1977	Ray et al	310/113 X
4,138,629	2/1979	Miller et al.	310/261 X
4,237,395	12/1980	Loudermilk	318/140
4,259,604	3/1981	Aoki	310/113
4,297,604	10/1981	Tawse	310/168

Primary Examiner—William M. Shoop, Jr.
Assistant Examiner—Shik Luen Paul Ip
Attorney, Agent, or Firm—Jim Zegeer

[57]   ABSTRACT

A unitized (single unit) motor and flux switch alternator having stationary field, armature and motor windings which provides a magnetic path for some of the motor input power to feed through and increase the alternating current (AC) generator output. A rotor formed from a material having a high magnetic permeability (solid or laminated soft steel) is controlled in speed by controlling the magnitude and timing of the pulsed direct current (DC) supplied to the motor windings which may be wound on the stationary legs or the rotor. The current flow in the motor windings can be controlled by a mechanical commutator if the motor windings are on the rotor or by a solid-state converter if the motor windings are on the legs in a manner normally associated with brushless DC motors. The DC windings of the flux switch alternator can be replaced by permanent magnets since the reversing field in the AC output windings are predominantly time stationary.

8 Claims, 7 Drawing Figures

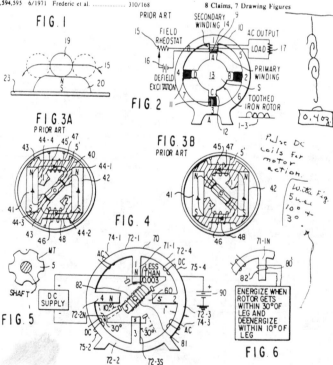

BACKGROUND OF THE INVENTION

Inductor alternators were as popular and efficient as any generator before 1900. They had no brushes, no high reliability but they were slightly larger than other generators and output unidirectional pulses. As a result they lost out to other generators except in special applications. Later the flux switch alternator replaced the inductor alternator as the flux switch alternator outputs AC and since all AC coils and DC coils were used twice as much, the flux switch alternator output four times more than inductor alternator, all else being equal.

Simple inductor alternators had four legs with AC and DC coils wound on each leg and a four lobed steel rotor. The flux switch alternator simply wound these same coils between the four legs instead of on the legs and cut two opposite lobes from the steel rotor. Since only steel rotates with a conservative force, what could require four times more input torque to the flux switch alternator?

Because of sags, glitches, brownouts, and other surprises from electric power systems many large electronic systems including computers now use a motorgenerator (M-G) for back-up or emergency power. Few motors or generators are individually over 95 percent efficient so when their shafts are mechanically coupled, the overall efficiency of an M-G with separate motors and generators is seldom over 90 percent efficient.

It is commonplace to teach the output of a generator is equal to the mechanical input power minus the losses. It is also known from Lenz's law (but seldom taught) a generator that is 95 percent efficient consumes 95 percent of the input to overcome torque due to internal forces and 5 percent goes to losses. The rotors of most of today's generators are repelled as they approach a stator and are attracted back by the stator as soon as the rotor passes the stator in accordance with Lenz's law. Thus, most rotors face constant nonconservative work forces and therefore, present generators require constant input torque.

Therefore, it is an object of this invention to provide a more compact motor generator.

It is also an objective of this invention to bias all steel above ground by attaching this steel to the positive terminal of a power supply or battery and grounding the negative terminal to bleed off or ground most free electrons to decrease losses from unwanted induced currents. This will also decrease losses in any other motor, generator or transformer with armatures.

It is further an objective of this invention to make a more compact and far more-efficient motor generator by unitization.

It is yet another objective of this invention to take advantage of a conservative no work force demonstrated by a simple damped oscillator consisting of a steel ball bearing released off center on a button permanent magnet with magnetic poles on the flat surfaces.

According to this invention, the legs or the rotor of a flux switch alternator are provided with motor windings. The steel rotor of the unitized flux switch alternator actually aids the input torque for half of each rotation as the rotor is always attracted and never repelled. This construction makes it possible for some of the current or power fed to the motor windings to magnetically feed through a solid magnetic path to the AC output windings which does not occur in today's M-Gs as they are only mechanically coupled by their shafts and have no common magnetic path to share.

From basic electronic technology we learn a charged condensor has few free or conduction electrons on the positive plate and an excess of free electrons on the negative or grounded plate. Since steel armatures are conductors, there has been considerable effort expended in materials research to increase resistance to conduction electrons in armature materials to thereby reduce hysteresis and eddy current damping losses. Another more common approach is to laminate or powder these armatures. Accordingly, a further feature of the invention, the reduction in hysteresis and eddy current damping losses.

United States Patent

Ecklin

3,879,622

Apr. 22, 1975

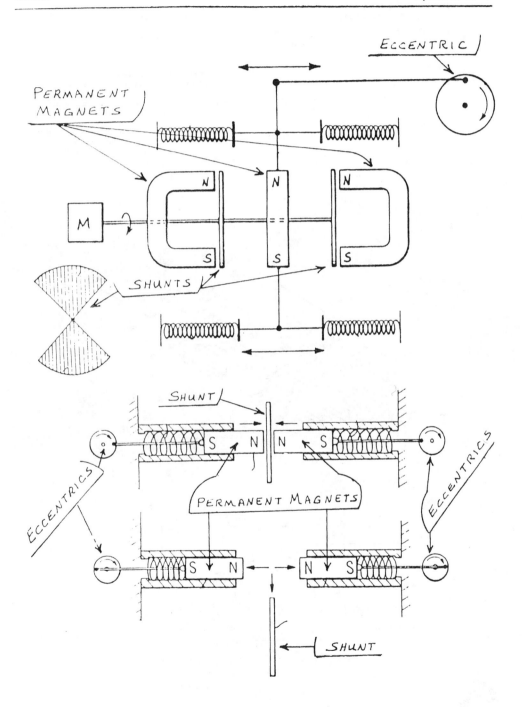

United States Patent [19]

Alexander

[11] **3,913,004**

[45] **Oct. 14, 1975**

[54] **METHOD AND APPARATUS FOR INCREASING ELECTRICAL POWER**

[75] Inventor: Robert W. Alexander, Pasadena, Calif.

[73] Assignee: Alex, Pasadena, Calif.

[22] Filed: Nov. 18, 1974

[21] Appl. No.: 524,556

[52] U.S. Cl. 321/28; 321/50
[51] Int. Cl.² H02M 7/64
[58] Field of Search 310/113, 165; 321/28, 29, 321/30, 31, 48, 49, 50

[56] **References Cited**
UNITED STATES PATENTS

2,640,181 5/1953 Korzdorfer........................ 321/28 X

3,078,409 2/1963 Bertsche, Jr. et al............. 321/28 X
3,223,916 12/1965 Shafranek et al..................... 321/28

Primary Examiner—William M. Shoop

[57] **ABSTRACT**

A form of rotating machine arranged in such a way as to convert a substantially constant input voltage into a substantially constant output voltage; involving generally a rotor that revolves at substantially constant speed within a stator and which comprises a transformer core subjected to and having a primary motor-transformer winding and a secondary transformer-generator winding; whereby transformed and generated power are synchronously combined as increased output power.

27 Claims, 3 Drawing Figures

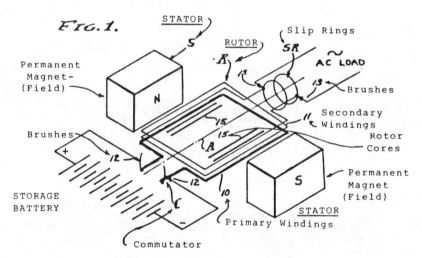

FIG. 1.

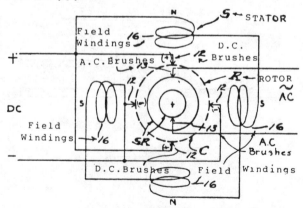

FIG. 2.

United States Patent [19]

Gray

[11]   **3,890,548**

[45] **June 17, 1975**

[54] **PULSED CAPACITOR DISCHARGE ELECTRIC ENGINE**

[75] Inventor: Edwin V. Gray, Northridge. Calif.

[73] Assignee: Evgray Enterprises, Inc., Van Nuys, Calif.

[22] Filed: Nov. 2, 1973

[21] Appl. No.: 412,415

[52] U.S. Cl. 318/139; 318/254; 318/439; 310/46

[51] Int. Cl. H02p 5/00

[58] Field of Search 310/46, 5, 6; 318/194, 318/439, 254, 139; 320/1; 307/110

[56]   **References Cited**

UNITED STATES PATENTS

2,085,708	6/1937	Spencer	318/194
2,800,619	7/1957	Brunt	318/194
3,579,074	5/1971	Roberts	320/1
3,619,638	11/1971	Phinney	307/110

OTHER PUBLICATIONS

Frungel, *High Speed Pulse Technology*, Academic Press Inc., 1965, pp. 140–148.

Primary Examiner—Robert K. Schaefer
Assistant Examiner—John J. Feldhaus
Attorney, Agent, or Firm—Gerald L. Price

[57]   **ABSTRACT**

There is disclosed herein an electric machine or engine in which a rotor cage having an array of electromagnets is rotatable in an array of electromagnets, or fixed electromagnets are juxtaposed against movable ones. The coils of the electromagnets are connected in the discharge path of capacitors charged to relatively high voltage and discharged through the electromagnetic coils when selected rotor and stator elements are in alignment, or when the fixed electromagnets and movable electromagnets are juxtaposed. The discharge occurs across spark gaps disclosed in alignment with respect to the desired juxtaposition of the selected movable and stationary electromagnets. The capacitor discharges occur simultaneously through juxtaposed stationary movable electromagnets wound so that their respective cores are in magnetic repulsion polarity, thus resulting in the forced motion of movable electromagnetic elements away from the juxtaposed stationary electromagnetic elements at the discharge, thereby achieving motion. In an engine, the discharges occur successively across selected ones of the gaps to maintain continuous rotation. Capacitors are recharged between successive alignment positions of particular rotor and stator electromagnets of the engine.

18 Claims, 19 Drawing Figures

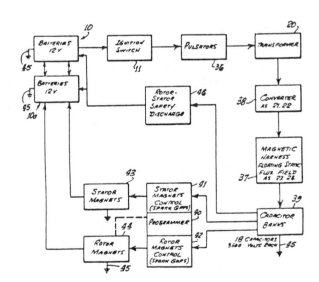

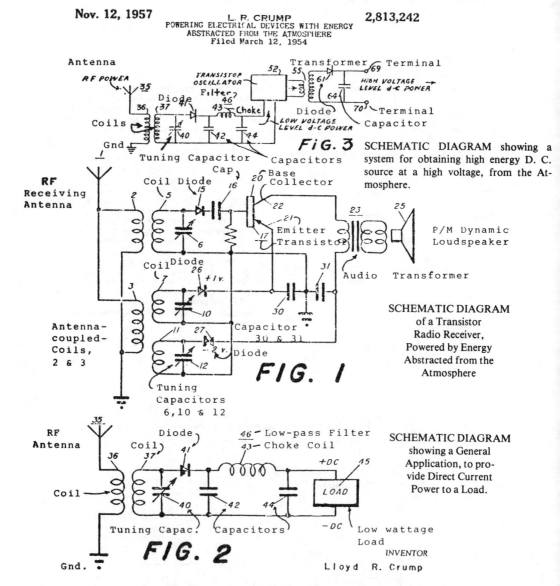

Nov. 12, 1957

L. R. CRUMP

2,813,242

POWERING ELECTRICAL DEVICES WITH ENERGY
ABSTRACTED FROM THE ATMOSPHERE
Filed March 12, 1954

FIG. 3 SCHEMATIC DIAGRAM showing a
system for obtaining high energy D. C.
source at a high voltage, from the Atmosphere.

SCHEMATIC DIAGRAM
of a Transistor
Radio Receiver,
Powered by Energy
Abstracted from the
Atmosphere

SCHEMATIC DIAGRAM
showing a General
Application, to pro-
vide Direct Current
Power to a Load.

INVENTOR

Lloyd R. Crump

BY *W. E. Thibodeaux & A. M. Dew*

ATTORNEYS

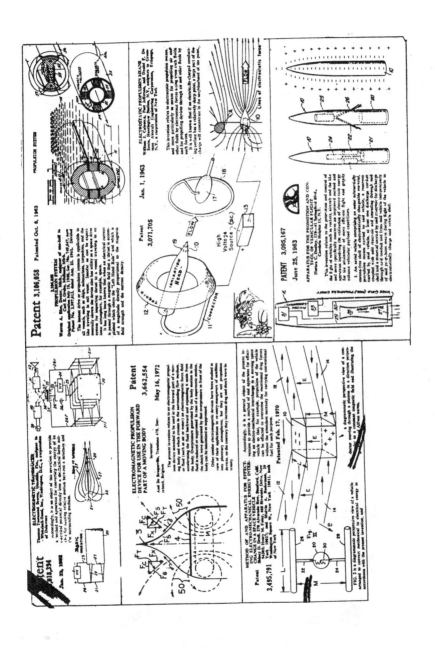

Bill Muller   PRAN Technologies, Penticton, Canada

Bill Muller and his group have evolved a unique and very promising PM/EM unit. The design is based on using an even number of rotor segments and uneven number of iron stator segments, unlike conventional generators which have matched, even sets of segments.

The even number of rotor segments are permanent magnets, while the fixed, uneven number of iron stator segments provide an overlapping unbalanced alternate attraction and repulsion between the opposite interacting segments. Since the electrical coils are directly associated with each iron stator segments, this portion of the design is conventional, but the timed capacitive discharge into the coils makes this design arrangment unconventional. The capacitive discharge principle has been proven by Ed Gray, with his EvGray unit, as described in Group 8.

In addition to the natural advantage of utilizing the odd-even numbers of interacting rotor-stator segments,   s design avoids the disadvantages of conventional generators by the elimination of back-EMF, ie: Lenz's Law, with the capacitive-discharge input into the multiple stator coils.

There are some key design points which must be strictly followed in order for these types of units to work properly:-

a)The permanent magnets used must be of the SAMCO-(samarium/cobalt or NIB magnets.

b)The iron segment - (rotor component) must be of equal width to the width of the P/M's used (stator component).

c)The spacing between the stator/field permanent magnets must not exceed 1.2 times the permanent magnet's width, at the rotating line.

d)The air gap between the rotor iron segments and the stationary field P/M's must be quite close at .010 and no greater than .020 in order to maximize the magnetic instability between the stator/rotor components.

It must be noted that Bill Muller's unique and practical motor is not just theory, there is at least one operating prototype which has been validated by a local Engineering Test Laboratory and witnessed by Drs. Petermannn and Schaffranke, who endorsed this new type P.M.M.

These types of permanent magnet motors are not self-starting and require some type of capacitive-discharge of input - (Ev-Gray) or an electromagnetic repulse similar to the Japanese Kure-Tekko design. Bill Muller states that this electromagnetic repulsing, or capacitive-discharge type of starting input can be made continuous as a running/timed function which would probably increase the effectiveness of these P.M.M's. As a continuous/running function this Muller design would then become a hybrid PM/EM unit, as described in Section 10(PM/EM motor).

Heat Pump Concept:-

Bill Muller has also developed a very useful heat pump arrangement which is a spinoff from the P.M.M. unit. In this unique concept, the permanent magnets are uniformly placed within the rotor in a multiple concentric pattern. The same "odd"/"even" method is used, with iron or "Metglass" segments placed within a similar multiple concentric pattern. Spiral piping passes through each of the iron or "Metglass" discs.

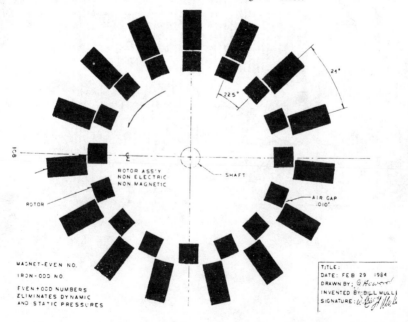

MAGNET-EVEN NO.

IRON-ODD NO.

EVEN+ODD NUMBERS
ELIMINATES DYNAMIC
AND STATIC PRESSURES

| TITLE: |
| DATE: FEB 29 1984 |
| DRAWN BY: |
| INVENTED BY BILL MULLER |
| SIGNATURE: |

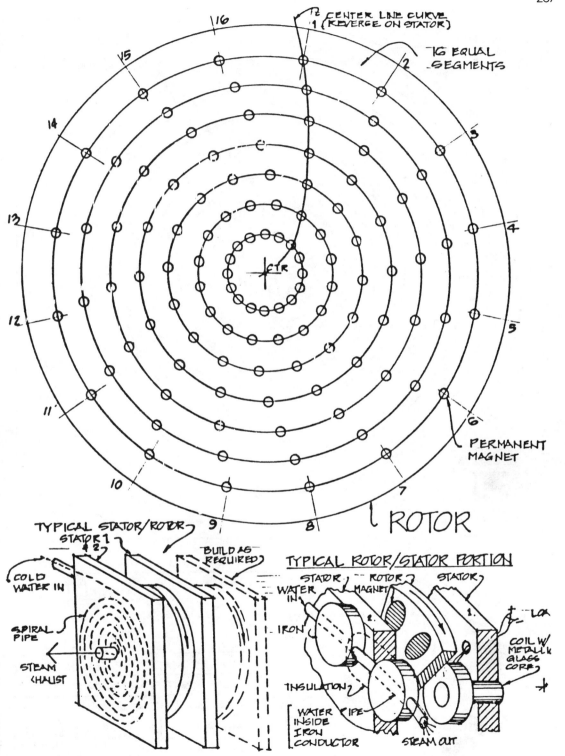

CENTER LINE CURVE
(REVERSE ON STATOR)

16 EQUAL
SEGMENTS

CTR

PERMANENT
MAGNET

ROTOR

TYPICAL STATOR/ROTOR

STATOR 1
(BUILD AS REQUIRED)

COLD WATER IN

SPIRAL PIPE

STEAM EXHAUST

TYPICAL ROTOR/STATOR PORTION

STATOR   ROTOR   STATOR
WATER IN   MAGNET

IRON

COIL W/ METALLIC GLASS CORE

INSULATION

WATER PIPE INSIDE IRON CONDUCTOR

STEAM OUT

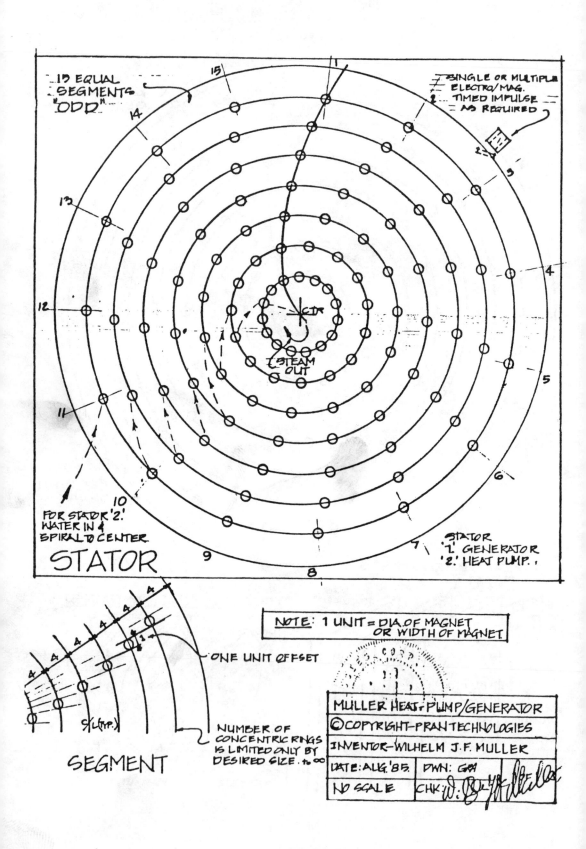

15 EQUAL
SEGMENTS
"ODD"

SINGLE OR MULTIPLE
ELECTRO/MAG.
TIMED IMPULSE
AS REQUIRED

CTR

STEAM
OUT

FOR STATOR '2'
WATER IN &
SPIRAL TO CENTER

STATOR

STATOR
'1.' GENERATOR
'2.' HEAT PUMP.

ONE UNIT OFFSET

S/(TYP.)

SEGMENT

NUMBER OF
CONCENTRIC RINGS
IS LIMITED ONLY BY
DESIRED SIZE. to ∞

NOTE: 1 UNIT = DIA. OF MAGNET
OR WIDTH OF MAGNET

MULLER HEAT PUMP/GENERATOR		
©COPYRIGHT-PRAN TECHNOLOGIES		
INVENTOR-WILHELM J.F. MULLER		
DATE: AUG.'85.	DWN: GM	
NO SCALE	CHK:	

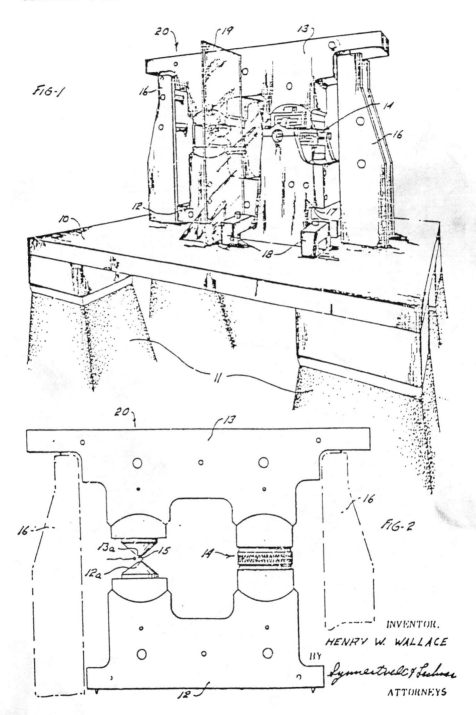

FIG-1

FIG-2

INVENTOR.
HENRY W. WALLACE
BY
ATTORNEYS

Dec. 14, 1971   H. W. WALLACE   3,626,605

METHOD AND APPARATUS FOR GENERATING A DYNAMIC FORCE FIELD

Filed Nov. 4, 1968   4 Sheets-Sheet 2

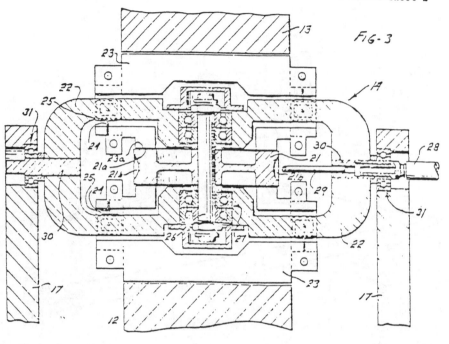

FIG-3

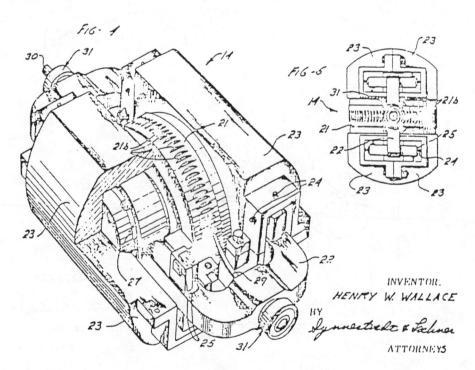

FIG- 1

FIG-5

INVENTOR.
HENRY W. WALLACE
BY
Lynnestedt & Lechner
ATTORNEYS

Dec. 14, 1971   H. W. WALLACE   3,626,606

METHOD AND APPARATUS FOR GENERATING A DYNAMIC FORCE FIELD

Filed Nov. 4, 1968   4 Sheets—Sheet 3

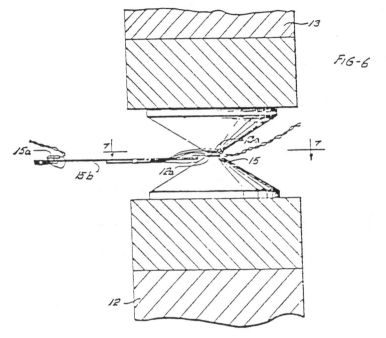

FIG-6

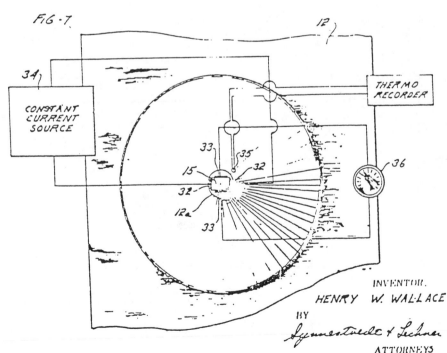

FIG·7

CONSTANT CURRENT SOURCE

THERMO RECORDER

INVENTOR.
HENRY W. WALLACE
BY
ATTORNEYS

262

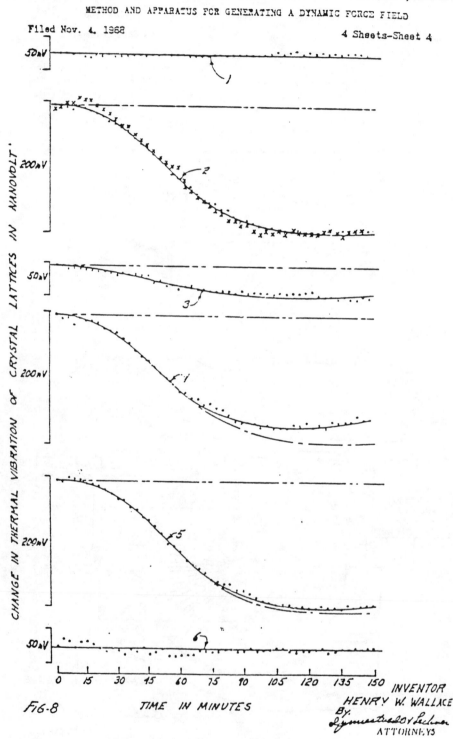

FIG-8   TIME IN MINUTES

INVENTOR
HENRY W. WALLACE
BY:
ATTORNEYS

1

3,626,606
METHOD AND APPARATUS FOR GENERATING A DYNAMIC FORCE FIELD
Henry W. Wallace, Ardmore, Pa.
(303 Cherry Lane, Laurel, Miss. 39440)
Filed Nov. 4, 1968, Ser. No. 773,116
Int. Cl. G09b 23/06
U.S. Cl. 35—19   10 Claims

ABSTRACT OF THE DISCLOSURE

Apparatus and method for generating a non-electromagnetic force field due to the dynamic interaction of relatively moving bodies through gravitational coupling, and for transforming such force fields into energy for doing useful work.

The method of generating such non-electromagnetic forces includes the steps of juxtaposing in field series relationship a stationary member, comprising spin nuclei material further characterized by a half integral spin value, and a member capable of assuming relative motion with respect to said stationary member and also characterized by spin nuclei material of one-half integral spin value; and initiating the relative motion of said one member with respect to the other, whereby the interaction of the angular momentum property of spin nuclei with inertial space effects the polarization of the spin nuclei thereof, resulting in turn in a net component of angular momentum which exhibits itself in the form of a dipole moment capable of dynamically interacting with the spin nuclei material of the stationary member, thereby further polarizing the spin nuclei material in said stationary member and resulting in a usable non-electromagnetic force.

This invention relates to an apparatus and method for use in generating energy arising through the relative motion of moving bodies and for transforming such generated energy into useful work. In the practice of the present invention it has been found that when bodies composed of certain material are placed in relative motion with respect to one another there is generated an energy field therein not heretofore observed. This field is not electromagnetic in nature; being by theoretical prediction related to the gravitational coupling of moving bodies.

The initial evidence indicates that this nonelectromagnetic field is generated as a result of the relative motion of bodies constituted of elements whose nuclei are characterized by half integral "spin" values; the spin of the nuclei being synonymous with the angular momentum of the nucleus thereof. The nucleons in turn comprise the elemental particles of the nucleus; i.e., the neutrons and protons. For purposes of the present invention, the field generated by the relative motion of materials characterized by a half integral spin value is referred to as a "kinemassic" force field.

It will be appreciated that relative motion occurs on various levels, i.e., there may be relative motion of discrete bodies as well as of the constituents thereof including, on a subatomic level, the nucleons of the nucleus. The kinemassic force field under consideration is a result of such relative motion, being a function of the dynamic interaction of two relatively moving bodies including the elemental particles thereof. The value of the kinemassic force field, created by reason of the dynamic interaction of the bodies experiencing relative motion, is the algebraic sum of the fields created by reason of the dynamic interaction of both elementary particles and of the discrete bodies.

For a closed system comprising only a stationary body, the kinemassic force, due to the dynamic interaction of

2

the particles therein, is zero because of the random distribution of spin orientations of the respective particles. Polarization of the spin components so as to align a majority thereof in a preferred direction establishes a field gradient normal to the spin axis of the elementary particles. The present invention is concerned with an apparatus for establishing such a preferred orientation and as a result generating a net force component capable of being represented in various useful forms.

According, the primary object of the present invention concerns the provision of means for generating a kinemassic field due to the dynamic interaction of relatively moving bodies.

A further object of the present invention concerns a force field generating apparatus wherein means are provided for polarizing material portions of the apparatus so as to reorient the spin of the elementary nuclear components thereof in a preferred direction thereby generating a detectable force field.

The kinemassic force field finds theoretical support in the laws of physics, being substantiated by the generalized theory of relativity. According to the general theory of relativity there exists not only a static gravitational field but also a dynamic component thereof due to the gravitational coupling of relatively moving bodies. This theory purposes that two spinning bodies will exert force on each other. Heretofore the theoretical predictions have never been experimentally substantiated however, as early as 1896, experiments were conducted in an effort to detect predicted centrifugal forces on stationary bodies placed near large, rapidly rotating masses. The results of these early experiments were inconclusive, and little else in the nature of this type of work is known to have been conducted.

It is therefore another object of the present invention to set forth an operative technique for generating a measurable force field due to gravitational coupling of relatively moving bodies.

Another more specific object of the present invention concerns a method of generating a non-electromagnetic force field due to the dynamic interaction of relatively moving bodies and for utilizing such forces for temperature control purposes including the specific application of such forces to the control of lattice vibrations within a crystalline structure thereby establishing an appreciable temperature reduction, these principles being useful for example in the design of a heat pump.

The foregoing objects and features of novelty which characterize the present invention as well as other objects of the invention are pointed out with particularity in the claims annexed to and forming a part of the present spection. For a better understanding of the invention, its advantages and specific objects allied with its use, reference should be made to the accompanying drawings and descriptive matter in which there is illustrated and described a preferred embodiment of the invention.

In the drawings:

FIG. 1 is an overall perspective view of equipment constructed according to the present invention, this equipment being designed especially for demonstrating the useful applications of kinemassic force fields;

FIG. 2 is an isolation schematic of apparatus components comprising the kinemassic field circuit of the apparatus of FIG. 1, showing the field series relationship of generator and detector units;

FIGS. 3, 4 and 5 show the generator of FIGS. 1 and 2 in greater detail;

FIG. 6 is an enlarged view of the detector working air gap area of the apparatus of FIGS. 1 and 2;

FIG. 7 is a sectional view of FIG. 6 showing associated control and monitoring equipment; and

FIG. 8 represents measured changes in the operating characteristics of a crystalline target subject to a kinemassic force field generated in the apparatus of FIGS. 1 and 2.

Before getting into a detailed discussion of the apparatus and steps involved in the practice of the present invention it should be helpful to an understanding of the present invention if consideration is first given to certain defining characteristics many of which bear an analogous relationship to electromagnetic field theory. A first feature is that the kinemassic field is vectorial in nature. The direction of the field vector is a function of the geometry in which the relative motion between mass particles takes place.

The second significant property of the kinemassic field relates the field strength to the nature of the material in the field. This property may be thought of as the kinemassic permeability by analogy to the concept of permeability in magnetic field theory. The field strength is apparently a function of the density of the spin nuclei material comprising the field circuit members. Whereas the permeability in magnetic field theory is a function of the density of unpaired electrons, the kinemassic permeability is a function of the density of spin nuclei and the measure of magnitude of their half integral spin values. As a consequence of this latter property, the field may be directed and confined by interposing into it denser portions of desired configuration. For example, the field may be in large measure confined to a closed loop of dense material starting and terminating adjacent a system wherein relative motion between masses is occuring.

A further property of the kinemassic force field relates field strength to the relative spacing between two masses in relative motion with respect to one another. Thus, the strength of the resultant field is a function of the proximity of the relatively moving bodies such that relative motion occurring between two masses which are closely adjacent will result in the generation of a field stronger than that created when the same two relatively moving bodies are spaced farther apart.

As mentioned above, a material consideration in generating the kinemassic force field concerns the use of spin nuclei material. By spin nuclei material is meant materials in nature which exhibit a nuclear external angular momentum component. This includes both the intrinsic spin of the unpaired nucleon as well as that due to the orbital motion of these nucleons.

Since the dynamic interaction field arising through gravitational coupling is a function of both the mass and proximity of two relatively moving bodies, then the resultant force field is predictably maximized within the nucleus of an atom due to the relatively high densities of the nucleons, both in terms of mass and relative spacing, plus the fact that the nucleons possess both intrinsic and orbital components of angular momentum. Such force fields may in fact account for a significant portion of the nuclear binding force found in all of nature.

It has been found that for certain materials, namely those characterized in a half integral spin value, the external component of angular momentum thereof will be accompanied by a force due to the dynamic interaction of the nucleons. This is the so-called kinemassic force which on a submacroscopic basis exhibits itself as a field dipole moment aligned with the external angular momentum vector. These moments are of sufficient magnitude that they interact with adjacent, or near adjacent spin nuclei field dipole moments of neighboring atoms.

This latter feature gives rise to a further analogy to electromagnetic field theory in that the interaction of adjacent spin nuclei field dipole moments gives rise to nuclear domain-like structures within matter containing sufficient spin nuclei material.

Although certain analogies exist between the kinemassic force field and electromagnetic field theory, it should be remembered that the kinemassic force is essentially nonresponsive to or affected by electromagnetic force phenomena. This latter condition further substantiates the ability of the kinemassic field to penetrate through and extend outward beyond the ambient electromagnetic field established by the moving electrons in the atomic structure surrounding the respective spin nuclei.

As in electromagnetic field theory, in an unpolarized sample, the external components of angular momentum of the nuclei to be subjected to a kinemassic force field, are originally randomly oriented such that the material exhibits no residual kinemassic field of its own. However, establishing the necessary criteria for such a force field effects a polarization of the spin components of adjacent nuclei in a preferred direction thereby resulting in a force field which may be represented in terms of kinemassic field flux lines normal to the direction of spin.

The fact that spin nuclei material exhibits external kinemassic forces suggests that these forces should exhibit themselves on a macroscopic basis and thus be detectable, when arranged in a manner similar to that for demonstrating the Barnett effect when dealing with electromagnetic phenomena.

In the Barnett effect a long iron cylinder, when rotated at high speed about its longitudinal axis, was found to develop a measurable component of magnetization, the value of which was found to be proportional to the angular speed. The effect was attributed to the influence of the impressed rotation upon the revolving electronic systems due to the mass property of the unpaired electrons within the atoms.

In the apparatus constructed in accordance with the foregoing principles it was found that a rotating member composed of spin nuclei material exhibits a kinemassic force field. The interaction of the spin nuclei angular momentum with inertial space causes the spin nuclei axes of the respective nuclei of the material being spun to tend to reorient parallel with the axis of the rotating member. This results in the nuclear polarization of the spin nuclei material. With sufficient polarization, an appreciable field of summed dipole moments emanates from the wheel rim flange surfaces to form a secondary dynamic interaction with the dipole moments of spin nuclei contained within the facing surface of a stationary body positioned immediately adjacent the rotating member.

When the stationary body, composed of suitable spin nuclei material, is connected in spatial series with the rotating member, a circuitous form of kinemassic field is created; the flux of which is primarily restricted to the field circuit.

Having now further defined the substantiating theory giving rise to the kinemassic forces operative in the present invention, reference is now made to the aforementioned drawings depicting in general an apparatus embodying the defining characteristics outlined above.

From the foregoing discussion, it will be appreciated that for both the purpose of detecting and exploiting the kinemassic field, several basic apparatus elements are necessary. First, apparatus is needed to enable masses to be placed in relative motion to one another. In order to maximize field strength the apparatus should be capable of generating high velocities between the particles in relative motion. Furthermore, the apparatus should be configured so that the proximity of the particles which are in relative motion is maximized. The necessity of using relatively dense material comprising half integral spin nuclei for the field circuit has already been stressed. These and other features are discussed in greater detail below in explanation of the drawings depicting an implementation of the invention, primarily for detection of the kinemassic field.

In considering the drawings, reference will first be made to the general arrangement of components, as particularly shown in FIGS. 1 and 2. As viewed in FIG. 1, the equipment is mounted upon a stationary base comprising a horizontal structure element 10 which rests upon permanent pilings of poured concrete 11 or other suitable structurally rigid material. It should be

5

made clear at the outset that the stationary base although not a critical element in its present form nevertheless serves an important function in the subject invention. Thus, the stationary base acts as a stabilized support member for mounting the equipment and, perhaps more significantly, the horizontal portion thereof is of such material that it tends to localize the kinemassic force field to the kinemassic force field generating apparatus proper. This latter feature is discussed in more detail below. The surface uniformity of the horizontal structural element 10 also facilitates the alignment of equipment components. In the reduction to practice embodiment of the present invention a layer of shock absorbing material (not shown) was interposed beneath the stationary base and the floor.

Shown mounted on the horizontal structural element 10 is the kinemassic force field generating apparatus indicated generally as 20, the lower portion of which is referred to as the lower mass member 12. An upper mass member 13 is positioned in mirrored relationship with respect to member 12 and separated somewhat to provide two air gaps therebetween. The lower and upper mass members 12 and 13 function as field circuit members in relationship to a generator 14 and a detector 15 positioned within respective ones of said two gaps. The spatial relationship of the generator, the detector and the mass members is such as to form a kinemassic force field series circuit.

All of the material members of the field circuit are comprised of half integral spin material. For example the major portion of the generator 14, and the upper and lower mass members 13 and 12, respectively, are formed of a particular brass alloy containing 89% copper, of which both isotopes provide a three-halves proton spin, 10% zinc, and 1% lead, as well as traces of tin and nickel. The zinc atom possesses one spin nuclei isotope which is 4.11% in abundance and likewise the lead also contains one spin nuclei isotope which is 22.6% in abundance. In order to gain an estimate of apparatus size, the upper circuit member has an overall length of 56 centimeters and a mass of 43 kilograms.

It will be seen that the constituents of the mass members are such as satisfy the criteria of half integral spin nuclei material for those apparatus parts associated with the field and the use of non-spin nuclei material for those parts where it is desired to inhibit the field. Accordingly, all support or structural members, such as the horizontal structural element 10, consist of steel. The iron and carbon nuclei of these structural members are classed as no-spin nuclei and thus represent high relative reluctance to the kinemassic field. Supports 16 are provided to accommodate the suspension of the upper mass member 13. The supports 16 are of steel the same as the horizontal support element 10. The high relative reluctance of steel to the kinemassic field minimizes the field flux loss created in the field series circuit of mass members 12 and 13, the generator 14, and the detector 15. The loss of field strength is further minimized by employing high-reluctance isolation bridges at the points of contact between the lower and upper mass members 12 and 13, and the structural support members 10 and 16.

Shunt losses within the apparatus were, in general, minimized by employing the technique of minimum mass contact; the use of low field permeability material at the isolation bridges or structural connections; and avoiding bulk mass proximity.

A number of techniques were developed for optimizing the isolation bridge units including Carboloy cones and spherical spacers. As is depicted more clearly in FIGS. 3, 4 and 5, the structural connection unit ultimately utilized consisted of a hardened 60° steel cone mounted within a setscrew and bearing against a hardened steel platen. The contact diameter of the cone against the platen measured approximately 0.007 inch and was loaded within elastic

6

limits. Adjustment is made by means of turning the setscrew within a mated, threaded hole.

FIG. 2 is presented in rather diagrammatic form; however, the diagrammatic configuration emphasizes that it consists of a rotatable member corresponding to the generator 14 of FIG. 1 which is "sandwiched" between a pair of generally U-shaped members corresponding to the lower and upper mass members 12 and 13 of FIG. 1. The wheel of generator 14 is mounted for rotation about an axis lying in the plane of the drawing. When member 14 is rotated rapidly with respect to the U-shaped members 12 and 13, a kinemassic field is generated which is normal to the plane defined by the rotating member and within the plane of the drawing.

As such, it may be represented in the drawing of FIG. 2 as taking a generally counterclockwise direction with respect to the field series circuit members.

Referring once more to FIG.1, it is seen that support for the generator unit 14 is provided by way of a support assembly 17, also fabricated of steel components. The support assembly 17 is in turn clamped to the horizontal structural element 10 by way of bracket assemblies 18.

In the embodiment of the present invention depicted in FIGS. 1 and 2, the lower and upper mass members 12 and 13 are fashioned into conical sections terminating in conical pole faces 12a and 13a in the area of the detector 15. This configuration tends to maximize the flux density in this area.

For isolation purposes, a curtain of transparent plastic material 19 is positioned so as to geometrically bisect the detector portion of the field circuit from the generator portion thereof. The function of the transparent curtain is to provide a degree of thermal isolation between the generator and detector units. Although not actually shown in FIG. 2 the transparent curtain is of H configuration and forms a vertical plane normal to the plane of the drawing and symmetrically positioned with respect thereto.

Not shown in the drawings are a tunnel of transparent material and a film of flexible plastic material which surround the detector 15 and associated equipment and thus serve to further stabilize the temperature conditions, thereby diminishing the adverse effects due to thermal gradients.

Before proceeding with the explanation of the operation of the apparatus disclosed in FIGS. 1 and 2, a more detailed description of certain portions of the structure will be given.

FIGS. 3, 4 and 5 present the generator assembly 14 of FIGS. 1 and 2 in greater detail. In particular, these figures disclose the relationship between a freely rotatable wheel 21, a bearing frame 22, and a pair of pole pieces 23. The bearing frame 22 is of structural steel, and functions to spatially orient the three generator parts without shunting the generated field potential.

Positioning of the generator wheel 21 with respect to the cooperative faces of the pole pieces 23 is effected by way of the bearing frame upon which the generator wheel is mounted. In this respect the high-reluctance isolation bridges mentioned with respect to FIGS. 1 and 2 are herein shown as setscrews 24 which are adjustably positioned to cooperate with hardened steel platens 25. The setscrews 24 are mounted on the pole pieces 23 and are adjustably positioned with respect to steel platens 25 cemented to the bearing frame 22 so as to facilitate the centering of the generator wheel 21 with respect to the interface surfaces 23a of the pole pieces 23.

In the implementation of the present invention the air gap formed between the generator wheel rim flanges and the stationary pole pieces 23 was adjusted to a light-rub relationship when the wheel was slowly rotated; as such this separation was calculated to be 0.001 centimeter for a wheel spin rate of 28,000 revolutions per minute due to the resulting hoop tension. In the drawing of FIG. 3 the spacing between the pole pieces 23 and the generator

3,626,606

7

wheel rim flange has been greatly exaggerated to indicate that in fact such a spacing does exist.

The generator wheel 21 utilized in the implementation of the present invention has an 8.60 centimeter diameter and an axial rim dimension of 1.88 centimeters. The rim flange surfaces 21a which are those field emanating areas closely adjacent the surfaces 23a of the pole pieces 23, are each 29 6 square centimeters. The rim portion of the wheel has a volume of 55.7 cubic centimeters neglecting the rim turbine slots 21b.

The generator wheel 21 and an associated mounting shaft 26 are mounted on the bearing frame 22 by means of enclosed double sets of matched high speed bearings 27.

Compressed air or nitrogen is used to drive the generator wheel by means of gas impingement against turbine buckets 21b cut in the wheel rim. The compressed gas is supplied through the supply line 28 and emanates from the air jet tube 29. Rate of spin is sensed by light rays reflected from the rim. For this purpose every other quadrant on the rim surface was painted black. Accordingly, light directed at the rim of the wheel will be reflected by the unpainted quadrants into light-sensing cells associated with a rate-measuring circuit of conventional design. Since the rate-detecting means form no part of the present invention they have not been depicted in the actual drawing.

Shaft members 39 carry suitable bearing members 31 for rotatably mounting the generator assembly with respect to a second axis. The support assembly 17 of FIG. 1 is partially represented in FIG. 4, and, as noted above it provides the mounting means for positioning the generator assembly 14 with respect to the lower and upper mass members 12 and 13.

Before proceeding with an explanation of the operation of the generator assembly with respect to the apparatus of FIG. 1, reference is made to FIGS. 6 and 7 which disclose an enlarged view of the detector 15. The lower and upper mass members 12 and 13 are given a conical configuration so as to maximize kinemassic field densities in the area of the working air gap, within which the detector is positioned. FIG. 7 represents a sectional view taken across the working air gap, showing the projection of the conical section of the upper mass member upon the conical section of the lower mass member. Although symmetrical in shape, the projection of the conical surface of the upper mass member onto the corresponding surface of the lower mass member has been slightly reduced for purposes of illustration. In the subject apparatus the two conical brass pole faces 12a and 13a form a working air gap measuring 0.114 centimeter across. Each disc shape pole face measures 0.71 square centimeter in area.

The detector or probe 15 is of indium arsenide and is inserted in the detector air gap with a spacing from either pole face of 0.02 centimeter, the target thickness measuring 0.07 centimeter. Both indium and arsenic process 100% isotope abundance of half integral spin nuclei; arsenic nuclei consists of one isotope of three halves proton spin, while indium nuclei are of two isotopes, both being of nine-halves proton spin.

A second probe of similar semi-conductor material 15a is shown in FIG. 6 as being positioned in close proximity to the first detector. Both probes 15 and 15a are shown mounted on a boom 15b which is shock mounted by means not shown. Shock mounting of the components is important due to the relatively close spacing between the probe and conical pole faces. Lateral displacement of the second probe from the vicinity of the working air gap measured as 25-centimeters.

Although not critical to the overall theory of the present invention, the selection of a semi-conductor probe of the nature heretofore described and the effective results realized through the positioning of the probe 15 and the associated probe 15a with respect to the working air gap between the conical pole faces as well as the manner in which signals measured by the two probes

8

is correlated, are important to an understanding of the forces involved. In this respect it is important to realize that the first and second semi-conductor probes were differentially connected in terms of electrical output and are polarity-sensitive to magnetic field measurements. Together the two probes constitute a differential magnetic probe for an FW Bell Gaussmeter. As conventionally used, such probes provide a measure of the magnetic field intensity from both AC and DC sources, via the Hall effect. The Hall effect is a well known phenomenon whereby a potential gradient is developed in a direction transverse to the direction of current flow within a conductor when the conductor is positioned in a magnetic field. It should be clearly understood, however, that no magnetic field phenomenon is associated with the present invention. Thus the lateral voltages which are measured in the present arrangement are not Hall voltages. This statement is substantiated by the explanation which follows, clearly establishing the absence of any Hall voltage indicative of magnetic fields. In this respect, the two probes are differentially connected for magnetic field measurements to eliminate errors due to ambient magnetic field changes whereas they are additively connected for sensing changes in thermal vibration of crystal lattices. Although polarity-sensitive to the magnetic field, the differential magnetic probe is not polarity-sensitive to changes in thermal vibration of crystal lattices.

The fact that the probes are polarity-sensitive with respect to magnetic field but not with respect to the direction of crystal lattice vibrations means that when the probes are reversed with respect to polarity any discernible difference in the output readings might be attributed to a magnetic field induced into the system by the rotating wheel. Inasmuch as the field conductive portions of the apparatus are comprised predominately of brass which is a paramagnetic material, no appreciable magnetic field should be detected. This in fact corresponds to the actual results in that no measurable difference in magnetic flux was recorded when the polarity of the probes was changed. It is thus possible to realistically discount magnetic fields as influencing operating results.

As seen in FIG. 7, the detector 15 has associated therewith two pairs of contacts 32 and 33, the first of which represents current contacts connected in turn to a source of constant current 34 of conventional design. The second set of contacts 33 are voltage contacts connected to detect any potential gradient transverse to the direction of current flow within the detector. The meter 36 represents means for detecting such potential differences and may be in the form of a very sensitive galvanometer.

A thermocouple 35 is positioned in close proximity to the detector 15 to monitor the temperature thereof. Temperature differences, as recorded by the thermocouple 35, are used for purposes of providing correction figures to the test results. A similar thermocouple is used in conjunction with the second detector 15a, as well as with the upper mass member particularly in the area of the generator wheel. Thermocouples are used for temperature monitoring since the energy change of their conducting electrons, by which they sense temperature change, are not measurably affected by the kinemassic field.

Proceeding now to an explanation of the operation of the subject invention, it will be appreciated that in accordance with the theory of operation of the present apparatus when the generator wheel is made to spin at rates upwards of 10 or 20 thousand revolutions per minute, effective polarization of spin nuclei within the wheel structure gradually occurs. This polarization gradually gives rise to domain-like structures which continue to grow so as to extend their field dipole moment across the interface separating the rim 21 from the pole pieces 23. Secondary dynamic interactions of gravitational coupling between respective dipoles increase the

field flux lines around the apparatus field circuit, thus
resulting in ever increasing total nuclear polarization of
half integral spin nuclei.

The non-electromagnetic forces so generated within the
subject apparatus are directed to the working air gap
within which is positioned the semi-conductor probe 15.
Therein the kinemassic forces are constructively used
to reduce the vibrational degrees of freedom of the crystal
lattice structure of the semi-conductor probe resulting in
a change of its electrical conductivity property. More
specifically, the kinemassic field, due to the dynamic interaction of the gravitational coupling of the mass components of the wheel in relation to the stationary portions of the pole pieces in immediate proximity therewith,
is restricted to the relatively high permeability material
comprising the lower and upper mass members, and is
concentrated at the working air gap by means of the
conical pole pieces. Inserted in the air gap is the probe
of semi-conductor spin nuclei material.

Control circuitry connected to two of the four contacts on the semi-conductor probe 15 is designed to
maintain a constant current flow across these contacts.
At the same time the ambient temperature of the area
surrounding the equipment is permitted to increase. In
fact the increase in ambient temperature is initiated well
in advance of the initiation of rotation of the generator
wheel giving rise to the non-electromagnetic kinemassic
force field. The constant increase in temperature is meant
to mask out otherwise positive and negative temperature
variations resulting in a reduced signal-to-noise ratio of
measurement.

In light of the gradual and constant increase in temperature of both the equipment and ambient conditions
surrounding the equipment, it might be expected that the
thermal vibrations of crystal lattice of the semi-conductor
probe would likewise increase. In actuality, a measurable
decrease in crystal lattice vibrations is detected within
the semi-conductor probe. The actual measurements recorded are in terms of nanovolts of meter movement, and
correspond to a decrease in lateral voltages measured
across the semi-conductor probe. These values can only
be accounted for by an effective polarization of the spin
nuclei of the lattice structure due to the polarizing effects
of the applied kinemassic force field. The polarization
results in a change in the specific heat property of the
crystal material, which in turn reflects itself as an increase in electrical conductivity measurable by the
galvanometer.

Reference is now made to FIG. 8 which discloses in
graphical relationship the results achieved by various
test arrangements of the semi-conductor probe with respect to the subject apparatus.

In the interpretation of the graphical relationships
of FIG. 8 it should be understood that corrections for
temperature variations have already been applied. These
temperature corrections account for the heat applied to
the system, that generated within the apparatus due to
frictional heating, as well as that due to the change in
specific heat property of the apparatus principally the
brass members due to their relative bulk. The latter component represents a positive contribution to the ambient
temperature due to the decrease in degrees of freedom of
the crystal lattice structure of the spin nuclei material
when subjected to the kinemassic force field. The above
mentioned heat factors result in the increased temperature of the brass members of the apparatus; these increases
being monitored by way of the thermocouples positioned
in proximity to the kinemassic field generating apparatus,
member 35 of FIG. 7 being an example thereof.

Curve 1 of FIG. 8 represents a static test conducted
over a period of 150 minutes, values being recorded at
3 minute intervals which was standard procedure for
the entire test series. Information gathered in respect to
curve 1 was useful in determining compensating factors
for ambient temperature changes. In curve 1 as well as

each of the other curves of FIG. 3, the ordinate values
measure a level of thermal vibration, in nanovolts of
meter movement, of InAs lattice structure against time
in which ambient temperature change of the two probes
has been quantitatively compensated.

Curve 2 represents a portion of a standard test run,
the portion shown being the active portion of the curve,
i.e., that portion of the curve for which measurable results were recorded due to the spinning of the generator
wheel. Not included in curve 2 are measurements taken
during a 78 minute preenergizing thermal calibration
period typical of the initial portion of each test run conducted. The pre-energizing thermal calibration period is
effected in order to illustrate the ambient temperature
compensation of the probes and as such is similar to
that of the static test of curve 1.

The first 45 minutes of the indicated 150 minute test
period of curve 2 represents the time during which the
wheel was made to spin at a rate of 28,000 r.p.m. The
continuity of the negatively sloping curve prior to, during
and following the time interval of the wheel returning
to its no spin state, and somewhat subsequently (an in-
dication of a return toward thermal equilibrium per-
centage distribution of spin angular-moment) is consistent
with the explanation advanced above concerning the
force field generated due to the dynamic interaction of
relatively moving bodies. It should be noted with respect
to curve 2 that separate test runs conducted some six
weeks apart tend to corroborate the independent test
results. The results of the two separate tests are superimposed in curve 2. These two tests, in addition to being
spaced in time, were spaced many test runs apart. The
two test results further establish the repeatibility of the
operation.

The change in thermal vibration of the InAs crystal
lattice for the test run of curve 2 is approximately equivalent to an 11° centigrade reduction in probe temperature. This figure has been substantiated by computer
studies. The computer has also been used to statistically
analyze the test data and establish the probability of
error in terms of the information recorded. In this respect the results of the computerized study indicate a
probability of error of 1 in 1 billion. Since any ratio
in excess of 1 in 20 eliminates the probability of chance
occurrence, the results obtained in the present instance
should be above reproach.

In order to substantiate the distance dependency of the
gravitational coupling force due to the dynamic interaction of relatively moving bodies it was predicted that
increasing the separation between the generator rim
flange 21a and the cooperating surface of the pole pieces
23a should measurably reduce the results obtained. The
results obtained when this separation was increased to
0.006 centimeters appears in curve 3. A comparison of
these results with those of curve 2 seemingly substantiates the conclusion that upon widening the gap a lessen-
ing of the dynamic interaction due to gravitational coupling between the spinning wheel and the stationary pole
piece actually occurs.

The data of curve 4 was taken with the air gap separation of the wheel to pole piece established at 0.001
centimeter as in the arrangement of curve 2; however,
the duration of wheel spin was decreased from 45 minutes to 30 minutes. Curve 2 results are shown superimposed on the solid line of curve 4. The relative magnitudes
of curves 2 and 4, when so contrasted with their respective
wheel spin periods, would appear to indicate a degree of
half integral spin nuclei polarization saturation.

Curve 5 depicts the results achieved by way of a shunt
test wherein two lead bars were secured to the stationary
brass bodies of the generator assembly so as to measure
the effect of shunting the field at zones of maximum field
potential. As contrasted with the results of curve 2, superimposed thereon, a statistically as well as visually significant difference is associated with the experimental re-

11

sults which, realistically, may be attributed to the shunting effect. The statistical study mentioned above, substantiates the distinguishable nature of the data groups resulting in curves 2 and 5.

Curve 6 depicts the results of a test run in which the field permeability has been eliminated by removal from the test apparatus of the upper mass member and the two detector conical pole faces. The lower mass member has also been adjusted downward so as to rest on the horizontal structural element 10. At the same time the spatial relationship between the generator assembly and the two differentially connected probes was not altered. As may be observed from the curve 6, there occurred no change in thermal vibration of the InAs crystal lattice. The plot scatter observable during the 45 minutes wheel spin period is attributable to increased temperature gradients which developed between the probes and the respective thermocouples in the absence of the various field circuit member thermal masses.

Further experimental results are available to substantiate the heretofore stated conclusions concerning the operating characteristics of the subject apparatus. In this respect reference is made to the copending application of the present inventor entitled Method and Apparatus For Generating a Secondary Gravitational Force Field, filed Nov. 4, 1968 and bearing Ser. No. 773,051, the subject of which concerns an apparatus for establishing a time variant kinemassic force field.

It will be apparent from the foregoing description that there has been provided an apparatus for generating and transforming kinemassic forces due to a dynamic interaction field arising through gravitational coupling of relatively moving bodies. Although in its original application the kinemassic force has been applied to the reduction of thermal vibrations in the lattice structure of a crystal, it should be readily apparent that other more significant uses of these forces are contemplated. In this respect the principles of the present invention may well be applied to any system in which bodies are nonresponsive or only partially responsive to conventional forces such as electromagnetic force fields. Thus, the present invention should have particular applicability to the stabilization of plasma particles, pursuant to controlled thermal nuclear fusion, or in the governing of temperatures and thermal energies within matter.

While in accordance with the provisions of the statutes, there has been illustrated and described the best forms of the invention known, it will be apparent to those skilled in the art that changes may be made in the apparatus described without departing from the spirit of the invention as set forth in the appended claims and that in some cases, certain features of the invention may be used to advantage without a corresponding use of other features.

Having now described the invention, what is claimed as new and for which it is desired to secure Letters Patent is:

1. An energy generating and transforming apparatus comprising a first member, said first member further comprised of spin nuclei material and mounted so as to be freely rotatable about an axis located within said first member, at least one stationary member, said stationary member comprised of spin nuclei material and positioned immediately adjacent said first member, and means for effecting the rotation of said first member whereby it is effective in impressing a non-electromagnetic force onto said stationary member.

2. A method for generating a non-electromagnetic force field and for converting such force field into useful work comprising the steps of mounting a first member comprised of preferred material in a manner which enables said first member to assume a degree of relative motion with respect to a second member also comprised of preferred material, establishing a degree of relative motion between said first and said second members, and

12

sensing the resultant energy due to the dynamic interaction of the relatively moving members.

3. The method of claim 2 wherein the sensing further comprises the steps of positioning a member of preferred material within said non-electromagnetic force field and measuring the change in the physical characteristics thereof.

4. An apparatus comprising two U-shaped members of spin nuclei material, non-spin nuclei material means for positioning said U-shaped members in mirrored relationship with one another and separated by two gaps, means including a freely rotatable member of spin nuclei material mounted in one of said two gaps, means including a detector mounted in the other one of said two gaps, and means for effecting the rotation of said freely rotatable member whereby a non-electromagnetic force is impressed upon said detector.

5. The apparatus of claim 4 wherein the detector positioned within the second of said two gaps comprises a crystalline structure of spin nuclei material such that the non-electromagnetic force impressed upon said crystalline structure is effective in polarizing said spin nuclei material sufficiently to reduce the specific heat properties of the crystalline structure so as to effect a substantial increase in the temperature thereof.

6. An energy generating apparatus comprising a first member, a second member, and means for establishing relative motion between said first and second members whereby a non-electromagnetic force is generated within said first and second members due to the dynamic interaction of said relatively moving members.

7. An energy generating and transforming apparatus comprising a mass circuit constructed of spin nuclei material of half integral spin value, said mass circuit having two gaps therein, field generating means rotatably mounted in one of said mass circuit gaps, said field generator means further comprising a frame for rotatably mounting thereon a member comprising spin nuclei material of half integral spin value, the axis of rotation of said rotatable member lying in the plane of said mass circuit, a pair of pole pieces mounted on said frame, said pole pieces being disposed on said frame on opposite sides of said rotatable member, each pole piece presenting a generally circular face in close proximity to but spaced from a face of said rotatable member, said pole pieces being further configured to substantially fill the gap in said mass circuit, means for rotating the rotatable member of said field generator means at high velocity, and means mounted in the other gap of said mass circuit for detecting a field in said circuit.

8. An energy generating and transforming apparatus comprising: a mass circuit of dense material, and having two gaps therein, mounting means for said mass circuit, said mounting means having restricted contact area with said mass circuit, field generator means rotatably mounted in one of said mass circuit gaps; said generator means further comprising a frame, a rotatable member mounted on said frame for rotation, the axis of rotation of said rotatable member lying in the plane of said mass circuit throughout all relative positions of said frame, a pair of pole pieces mounted on said frame by mounting means establishing restricted contact area between each pole piece and said frame, said pole pieces being disposed on said frame on opposite sides of said rotatable member, each pole piece presenting a generally circular face in close proximity to but spaced from a face of said rotatable member, said pole pieces being further configured to substantially fill the gap in said mass circuit, means for rotating the rotatable member of said generator means at high velocity, and means mounted in the other gap of said mass circuit for demonstrating a change in physical characteristics within said gap region due to the field generated within said mass circuit.

9. The apparatus of claim 8, wherein said means mounted in the other gap of said mass circuit comprises a

13

member whose atomic structure is such that it is affected by said field generated within said mass circuit.

10. A method for controlling the temperature in a crystalline structure by subjecting the crystalline structure to non-electromagnetic forces capable of altering the specific heat properties thereof, including the steps of: connecting in field series relation a mass circuit constructed of dense spin nuclei material of half integral spin value, a field generator constructed essentially of spin nuclei material having a half integral spin value and rotatably mounted in one of said mass circuit gaps, and a crystalline structure also of spin nuclei material having a half integral spin value positioned in the other of said mass circuit gaps; initiating the rotation of said field generator whereby the external angular momentum of spin nuclei material within said rotating field generator interacts with inertial space to effect the polarization of the spin nuclei

14

thereof, resulting in turn in a net component of angular momentum which dynamically interacts with the spin nuclei material of the mass circuit thereby further polarizing the nuclei of the material therein; and concentrating the resultant field within said field series circuit onto said crystalline structure within the second of said mass circuit gaps whereby the spin nuclei material of said crystalline structure is sufficiently polarized to reduce the specific heat properties of the crystalline structure due to a reduction in degrees of freedom of the lattice vibrations of said crystalline structure thereby effecting a substantial temperature increase in the body thereof.

No references cited.

HARLAND S. SKOGQUIST, Primary Examiner

United States Patent [19]

Eastham

[11]   **4,013,906**

[45]   **Mar. 22, 1977**

[54]   **ELECTROMAGNETIC LEVITATION**

[75]   Inventor: John Frederick Eastham, Aberdeen, Scotland

[73]   Assignee: National Research Development Corporation, London, England

[22]   Filed: **May 12, 1975**

[21]   Appl. No.: **576,663**

[30]   Foreign Application Priority Data

   May 14, 1974   United Kingdom 21256/74
   Oct. 19, 1974   United Kingdom 46481/74

[52]   U.S. Cl. .... 310/13; 104/148 LM;
                                    104/148 MS
[51]   Int. Cl.² H02K 41/02
[58]   Field of Search 310/12–14;
           104/148 R, 148 MS, 148 LM, 148 SS

[56]   **References Cited**

UNITED STATES PATENTS

3,717,103	2/1973	Guderjahn	104/148 SS
3,770,995	11/1973	Eastham et al.	310/13
3,836,799	9/1974	Eastham et al.	310/13
3,850,108	11/1974	Winkle	104/148 MS

OTHER PUBLICATIONS

IEEE Trans., "Traction and Normal Forces in the Linear Induction Motor" OOI & White, vol. Pas-89, No. 4, Apr. 1970, pp. 638–645.

Primary Examiner—Donovan F. Duggan
Attorney, Agent, or Firm—Cameron, Kerkam, Sutton, Stowell & Stowell

[57]   **ABSTRACT**

In a combined linear motor and levitation arrangement, the primary comprises two longitudinally extending rows of transverse U-cores disposed side by side. The secondary comprises electrically conductive non-magnetic material backed by magnetic material. The magnetic material of the secondary is formed in two parts each of which is associated with a respective row of U-cores of the primary.

7 Claims, 2 Drawing Figures

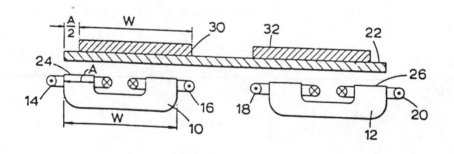

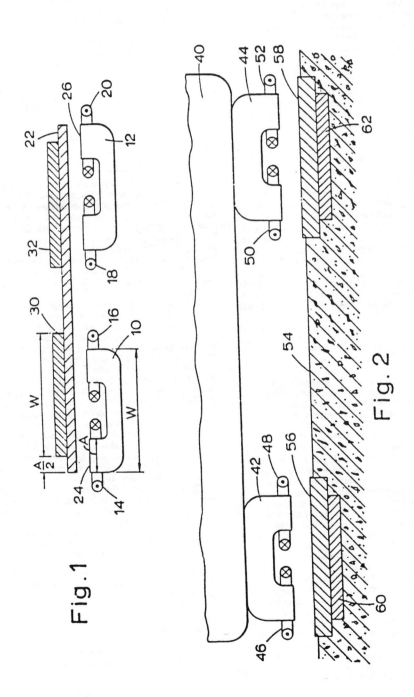

Fig.1

Fig.2

ELECTROMAGNETIC LEVITATION

This invention relates to electrical machines of the kind which employ electromagnetic levitation.

U.S. Pat. No. 3,836,799 relates to an electrical machine having a secondary comprising electrically conductive material and a primary comprising a core of magnetic material having at least two rows of pole faces confronting said secondary and being arranged to provide paths for working flux between said rows of pole faces in planes substantially perpendicular to said rows, the secondary being arranged to provide longitudinal paths for electric current on each side of each row of pole faces and transverse paths interconnecting said longitudinal paths, the primary being arranged, when energised from a polyphase alternating current supply, to create a magnetic field operative to produce a force between said primary and said secondary having first components tending to maintain said primary and said secondary spaced apart from one another, second components tending to maintain said primary and said secondary in alignment with one another in a direction perpendicular to said rows of pole faces and third components tending to cause relative displacement between said primary and said secondary in a direction parallel to said rows of pole faces. Thus this electrical machine consists of both a linear motor and a levitation arrangement.

It is already known that the use of magnetic material in the secondary of a levitation arrangement can improve the performance. For example, in the case of a simple coil floating (with no lateral stability) above a conducting sheet secondary, it has been shown that the levitated height per unit current can, in some circumstances be increased of the conducting sheet is backed by a sheet of steel. The dimensions of the system determine whether a height increase is obtained.

It is an object of the present invention to provide a system consisting of both a linear motor and a levitation arrangement having magnetic material in the secondary without impairing the lateral stability of levitation.

According to the present invention, there is provided an electrical machine having a primary comprising a core of magnetic material having at least two pairs of rows of pole faces disposed in a common plane and arranged to provide paths for working flux between the pole faces of each row in planes substantially perpendicular to said rows, and a secondary confronting said pole faces and comprising electrically conductive material arranged to provide longitudinal paths for electric current on each side of each row of pole faces and transverse paths interconnecting said longitudinal paths, and at least one core member of magnetic material extending longitudinally of the electrically conductive material on the opposite side thereof to the primary and having its lateral edges so disposed relative to the primary that the magnetic field produced when the primary is energized from a polyphase alternating current supply is operative to produce a force between said primary and said secondary having first components tending to maintain said primary and said secondary spaced apart from one another, second components tending to maintain said primary and said secondary in alignment with one another in a lateral direction and third components tending to cause relative displacement between said primary and said secondary in a longitudinal direction.

Preferably, the magnetic material of the secondary is in contact with the electrically conductive material of the secondary over two regions, one on each side of the longitudinal centre line of the electrically conductive material, the central region of the electrically conductive material not being in contact with the magnetic material. In one form of the invention, the secondary magnetic material consists of two members, one on each side of this central region.

In order that the invention may be more readily understood, embodiments thereof will now be described with reference to the accompanying drawing, in which:

FIG. 1 is a lateral cross-sectional view of a first embodiment of the invention, and

FIG. 2 is a part sectional, part end elevational view of a tracked ground transport system incorporating an electrical machine in accordance with a second embodiment of the invention.

Referring to FIG. 1, the primary of a levitation device in accordance with the invention comprises two longitudinally extending rows of U-shaped cores 10 and 12. Each core is oriented transversely and comprises a stack of laminations of magnetic material. Each core has a respective pair of windings such as the windings 14, 16, 18 and 20, one on each limb. The relative directions of current in the windings are as shown schematically in the drawing. Both rows of cores 10 and 12 have overall width W and have pole faces of width A as illustrated.

The secondary member comprises a sheet 22 of non-magnetic electrically conductive material such as aluminium. The width of the sheet of aluminium is substantially equal to the overall width of the primary core structure, i.e. from the outer edge of the limb 24 of the core 10 to the outer edge of the limb 26 of the core 12. Two secondary cores 30 and 32, of unlaminated magnetic material, are disposed on the opposite side of the sheet 22 to the primary cores 10 and 12. Both cores 30 and 32 have the same width W as the primary cores 10 and 12 but they are displaced inwardly relative to the primary cores 10 and 12 by an amount not substantially greater than the width A of the pole faces. Preferably this inward displacement is by a distance ¼A.

If the windings of successive U-shaped cores in each row are energized with single phase alternating current, a pure levitation effect is produced, the secondary being supported above the primary and stabilized laterally in alignment therewith. If successive coils in each row are energized phases of a polyphase alternating current supply, in addition to the levitation effect produced by the single phase energization, a travelling magnetic field is produced in the longitudinal direction so that the secondary now tends to move longitudinally with respect to the primary. As before, the secondary is supported above the primary and stabilized laterally. Any of the winding configurations described in the above-mentioned U.S. patent may be used.

Alternatively, the secondary cores 30 and 32 may be symmetrically disposed with respect to the corresponding primary cores 10 and 12 but, in this case, the lateral stability is decreased somewhat.

In FIG. 2, a vehicle 40 has a primary member similar to the primary of FIG. 1 in that it comprises two sets of primary cores 42 and 44 mounted on its underside. Each set of cores comprises two longitudinally extending rows of U-shaped cores, each core being oriented

April 28, 1964   A. P. DE SEVERSKY   3,130,945
                            IONOCRAFT

Filed Aug. 31, 1959   6 Sheets—Sheet 1

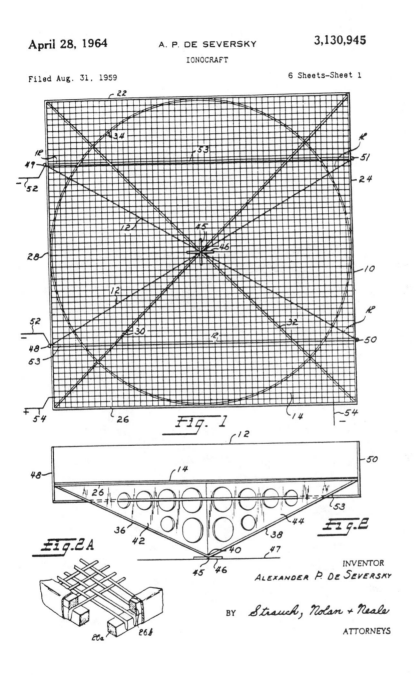

Fig. 1

Fig. 2

Fig. 2A

INVENTOR
ALEXANDER P. DE SEVERSKY

BY Strauch, Nolan + Neale

ATTORNEYS

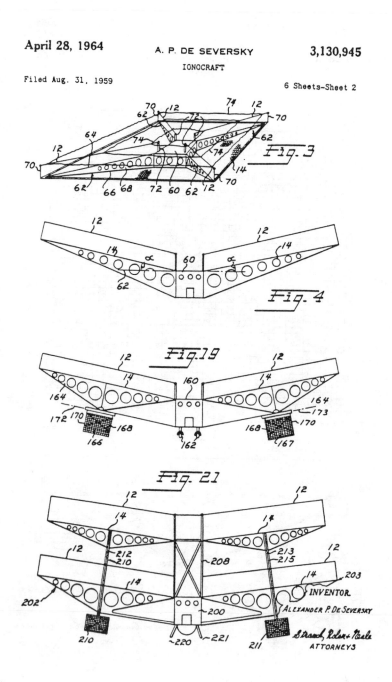

April 28, 1964   A. P. DE SEVERSKY   3,130,945
                                   IONOCRAFT

Filed Aug. 31, 1959   6 Sheets-Sheet 2

April 28, 1964   A. P. DE SEVERSKY   3,130,945
                                  IONOCRAFT

Filed Aug. 31, 1959   6 Sheets—Sheet 3

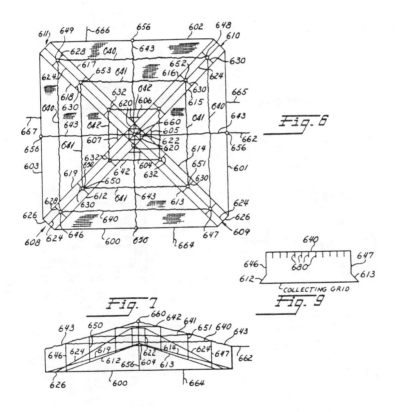

Fig. 6

Fig. 7

Fig. 9

COLLECTING GRID

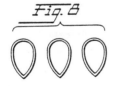

Fig. 8

INVENTOR
ALEXANDER P. DE SEVERSKY

BY Strauch, Nolan & Neale
                        ATTORNEYS

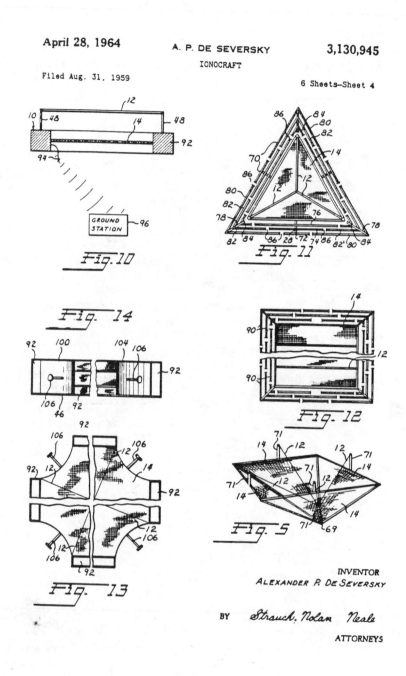

April 28, 1964   A. P. DE SEVERSKY   3,130,945
                           IONOCRAFT

Filed Aug. 31, 1959

                                    6 Sheets—Sheet 4

Fig. 10

Fig. 11

Fig. 14

Fig. 12

Fig. 13

Fig. 5

INVENTOR
ALEXANDER P. DE SEVERSKY

BY   Strauch, Nolan   Neale

ATTORNEYS

April 28, 1964   A. P. DE SEVERSKY   3,130,945
                              IONOCRAFT

Filed Aug. 31, 1959   6 Sheets—Sheet 5

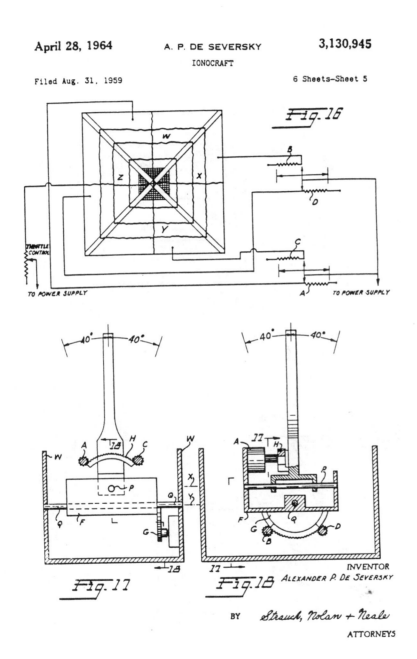

INVENTOR
ALEXANDER P. DE SEVERSKY

BY   Strauch, Nolan + Neale

ATTORNEYS

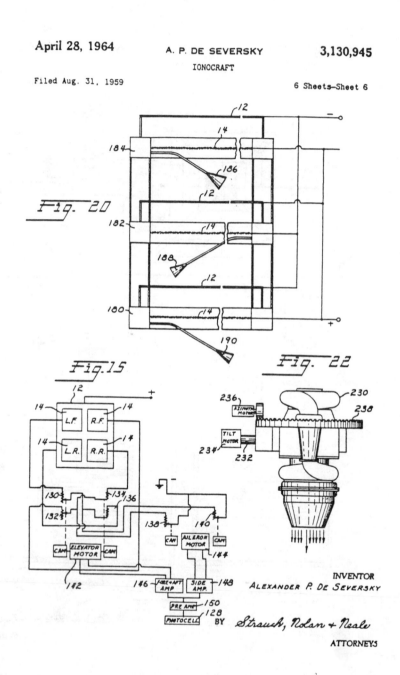

April 28, 1964   A. P. DE SEVERSKY   3,130,945
                            IONOCRAFT

Filed Aug. 31, 1959   6 Sheets—Sheet 6

Fig. 20

Fig. 15

Fig. 22

INVENTOR
ALEXANDER P. DE SEVERSKY

BY Strauch, Nolan & Neale

ATTORNEYS

United States Patent Office

3,130,945
Patented Apr. 28, 1964

1

2

3,130,945
IONOCRAFT
Alexander P. de Seversky, New York, N.Y., assignor to Electronatom Corporation, New York, N.Y., a corporation of New York
Filed Aug. 31, 1959, Ser. No. 837,150
29 Claims. (Cl. 244—62)

This invention relates to improved heavier-than-air aircraft, and more specifically to structures which are capable of either hovering or moving in any direction at high altitudes by means of ionic discharge.

The present invention is an improvement over well known electrostatic generation of winds used in a novel manner to supply propulsion and sustainance forces for a heavier-than-air aircraft. Crafts of the types herein disclosed having effective areas of several square feet have been successfully flown and contemplated platforms will inherently be of large size since the lift force is proportionate to the area through which large quantities or masses of air are accelerated downwardly from discharge electrodes to collection electrodes, the latter being a meshed-screen, bars, strips or any other structure that provides maximum collecting electrode area with perforations, slots or other types of opening to allow the air to pass through with a minimum of drag. Such a craft will be referred to in this application as an Ionocraft.

Such Ionocraft may serve as platforms which would be stationed above the earth for long periods of time and serve other purposes as will be explained below. The output power from microwave generators, such as magnetrons, coupled with high power capacity amplifier tubes may be beamed to the Ionocraft while airborne or the craft may carry its own power supply.

A principal object of the present invention is to provide a novel Ionocraft with space provided by the structure, preferably at the center of the craft, for installation of electronic equipment, and for the power plant, and crew where used.

Another object is to provide a novel Ionocraft construction wherein lightweight reinforcing members are provided to form a structure sufficiently rigid to cope with the dynamic and static loads and to maintain a desired distance between discharge emitting wires and the collecting grid.

Still another object resides in the novel configuration and arrangement of the emitting wires to assure uniform spacing from the collecting grid and to provide a maximum number of ionized particles for producing the desired lift.

A further object is to provide an improved Ionocraft of the foregoing type wherein some structural formation such as dihedral is provided for stabilizing the craft during flight. The dihedral may be positive or negative depending upon whether the hovering flight or horizontal motion of aircraft is a primary consideration of performance of the craft. A multiple deck structure may be used where desired to increase the lifting force, and dihedral may be provided in two or more angularly related directions to provide stability in all directions. A conical shape with the apex or nadir at the top or bottom center may also be advantageously used.

Still another object is to provide auxiliary ionic discharge structures mounted for rotational movement which are oriented to provide a horizontal propelling force and steering forces which can change the direction of the craft. By mounting such auxiliary structures to turn about a vertical axis, the craft can be made to turn in a horizontal plane about a vertical axis passing through the craft to thereby provide a scanning or target searching apparatus. A similar scanning motion can be achieved by mounting the auxiliary structures to turn about a horizontal axis.

A further object of this invention resides in the provision of a novel stick control using variable electrical impedances for control of the posture and for maneuvering the craft through variation of the voltage applied to different portions of the craft.

A second principal object of the present invention is to provide a combination Ionocraft and antenna system for radio frequency energy wherein the structure of the Ionocraft is so arranged as to serve in whole or in part as a structure of an efficient electromagnetic antenna system. In accordance with this object of the invention, the device contains one or more antennas that may be used for communication signal transmission, for detection, tracking and/or identification and for eventual destruction through collision of oncoming airborne or space vehicles or missiles and the like. The Ionocraft structure may be used, for example, as the main antenna element, as a series of directing or reflecting elements or as a parasitic element and may be shaped to provide arrays parabolas, corner reflectors, horns or lenses and be adapted to transmit a single or complete spectrum of frequencies from the extremely low frequencies to the highest frequencies including infrared.

Another object of this invention is to provide a combination antenna-Ionocraft with scanning means for detecting and/or tracking airborne vehicles or missiles. Such combination may also include suitable servo-control and other conventional equipment either on the Ionocraft or at a nearby ground station for causing the Ionocraft to "lock-on" automatically and/or be guided into the path of an "oncoming" vehicle or missile.

A further object is to provide an antenna which constantly locks on a radiation beam, such as a microwave or light beam for example, projected from the ground or from an aircraft in flight to change the position of the Ionocraft in flight.

These and other objects of the invention will become more fully apparent from the claims, and from the specification when read in conjunction with the appended drawings wherein:

FIGURES 1 and 2 (Sheet 1) are top plan and elevation views of the basic structure of an Ionocraft made in accordance with the present invention;

FIGURE 2a is an enlarged pictorial view of a portion of the structure showing how the grid wires are connected to the frame members;

FIGURE 3 (Sheet 2) is a pictorial view of a modified form of basic structure;

FIGURE 4 is a view in elevation of an embodiment similar to that shown in FIGURE 3 which is equipped with dihedral;

FIGURE 5 (Sheet 4) is a schematic view of a craft equipped with dihedral in two perpendicular directions;

FIGURES 6 and 7 (Sheet 3) are top plan and side elevation views respectively of a further embodiment of the present invention which is equipped with negative dihedral;

FIGURE 8 is a cross section of collecting grid structural members which may be used in lieu of the wire mesh;

FIGURE 9 is a view in elevation of an emitting wire having short wires suspended from the main wire to provide a point source for ion emission;

FIGURE 10 (Sheet 4) is a diagrammatic view in elevation of an Ionocraft in accordance with this invention;

FIGURES 11 and 12 are top plan views of two embodiments of the Ionocraft having a side elevation view as illustrated in FIGURE 10;

FIGURES 13 and 14 are top plan and elevation views of a further embodiment of this invention;

3,130,945

3

FIGURE 15 (Sheet 6) is a schematic diagram of a control circuit for causing the Ionocraft to lock-on and follow a radiation source at a ground station;

FIGURE 16 (Sheet 5) is a plan view, partly diagrammatic illustrating a control system for the craft of the present invention;

FIGURES 17 and 18 are side elevation views in section of a novel control stick box and assembly to permit steering and guiding of the craft by the system illustrated in FIGURE 16;

FIGURE 19 (Sheet 2) is a view in elevation of a craft having two vertical grid structure assemblies for controlling horizontal movement;

FIGURES 20 (Sheet 5) and 21 (Sheet 2) are side views in elevation of different embodiments each having several horizontal grid structures stacked one on top of the other: and

FIGURE 22 is a diagrammatic view of a gas turbine engine and mounting which are adapted for use with craft of the present invention.

Referring now to the drawings, FIGURES 1 and 2 (Sheet 1) are plan and elevation views of a typical basic embodiment of my improved Ionocraft 10. The Ionocraft proper comprises a plurality of emitting electrode wires 12 mounted above and in a plane substantially parallel to the collecting electrode grid 14 which may be composed of a meshed screen, bars, strips or any other structure that provides maximum effecting collecting electrode area with perforations, slots or other types of opening to allow the air to pass through with a minimum of drag. A plurality of hollow, lightweight rods or bars of conductive material or crossed wires forming a mesh which is open to pass air downwardly, but with the wires sufficiently closely spaced to effectively neutralize the charged ions which pass from emitting electrode wires 12 are preferred structures. A high D.C. voltage is applied between emitting electrode 12 and collecting electrode 14; one pole or terminal of the high voltage generator is connected to the emitting electrode 12 and the opposite pole or terminal of the same generator is connected to the collecting grid electrode 14, thus creating a high potential field between the electrodes.

In this form of improved Ionocraft, a basic structure sufficiently rigid to cope with the dynamic and static loads and to maintain a desired uniform distance between discharge emitting wires 12 and the collecting grid 14 is utilized comprising an outer square or rectangular frame composed of members 22, 24, 26 and 28. Diagonal frame members 30 and 32 extend between opposite corners of the rectangular frame and a circular frame member 34 is fixed tangentially to the midportions of the frame members. Said frame members are coplanar and collecting electrode wires 14 are interwoven, as with a loom, to form a closely meshed wire screen and supported from frame members 22, 24, 26 and 28. The ends of each wire are wrapped over and glued to the lower half 26a of the frame member and then cut off as shown in FIGURE 2A. The upper half 26b of the frame member is then secured in position as with glue. A considerable improvement in lifting force was achieved when the frame members and cut ends of the grid wires were covered with an aluminum foil.

Four lightweight rigid structural members, 36 and 38, of which two show in FIGURE 2, are mounted beneath the plane of collecting grid 14 in the vertical planes to diagonal members 30 and 32. Members 36 and 38 meet in a common junction 40 at the center of the Ionocraft. Four perforate lightweight rigid metal sheets or foils 42 and 44 of aluminum or the like, of which only two show in FIGURE 2, are mounted between diagonal members 30, 32, 36 and 38. These foils provide additional stabilization against tilting by guiding the air flow vertically along the surfaces of the foils and have been found to provide an increase in lift which more than compensates for their weight. Beneath junction 40, a pair of crossed

4

support members 45 and 46 are provided to serve as a landing support to hold the craft with the collecting grid 14 above the ground-supporting surface 47 when landed.

The outer ends of emitting wires 12 are supported from masts 48, 49, 50 and 51 of insulating material mounted on opposite sides of the craft. In this embodiment, emitting electrode wires 12 pass diagonally across the craft and cross each other near the center. One terminal of a high voltage D.C. potential is connected to leads 52 which are connected to masts 48 and 49.

The lower edges of masts 48 and 50 and of masts 49 and 51 are connected together by tension member 53 (FIGURE 2) such as a lightweight cable to hold the masts in their vertical position by providing a force to balance against the tension force of emitting wires 12.

Suitable lead-in wires 54 are provided for connection between collecting grid 14 and the other terminal of the power supply, and are preferably at ground potential. Variable impedances, such as variable width spark gaps which serve to reduce the applied voltage, are provided in lead-in wires 54 for control of voltage between emitting wires 12 and collecting grid 14 to thereby control the vertical movement of the craft.

An actual embodiment built in accordance with the foregoing description which lifted itself into a self-sustaining flight had a collecting grid surface area of approximately 150 square inches and the space between the collecting grid and the emitting wires 12 was approximately 2 inches. With a craft having the foregoing dimensions, a voltage of 20 kv. and a current of approximately 0.5 milliampere was sufficient to make the craft more than self-sustaining. The total weight of the structure was approximately 5 grams. Other craft having the space between the collecting grid and emitting wires of 5 inches have been successfully flown. Such craft require voltages of the order of 50 to 60 kv. Where the grid area is about 7 or 8 square feet, currents of the order of 2 milliamperes exist. Variations in humidity and air pressure cause variations in the current drawn and in the lifting efficiency.

The lifting capability of the craft was found to increase as the diameter of the grid wires is increased. Crafts were tested with wire diameters of 2, 5, 8 and 12 mils for the collecting grid. With wire diameters of 8 mils or more, the current requirement to provide the same total lifting force shows a detectable decrease thereby indicating a higher efficiency. Hollow tubular conductors having an outer diameter of one-quarter inch also give substantially the same lift force and efficiency as the 8 and 12 mil wire diameters.

A modification of the foregoing structure is shown in FIGURE 3 (Sheet 2) wherein a central compartment section 60 is provided in the center of a surrounding large area collecting grid 14. A plurality of rigid support sections 62, each comprising an upper member 64, a lower member 66, and an intermediate foil 68 extend from the corners of the central section 60 to the periphery of the framework surrounding the collecting grid 14.

Near each of the corners of the outer periphery of collecting grid 14 a mast 70 made of insulating material is provided which supports the outer end of emitting wires 12. A second group of inner support masts 72 mounted on central section 60 provide support for the inner ends of emitting wires 12.

In this embodiment, the central compartment 60 is adapted to house electronic equipment and the power plant and crew where used.

In practice, it has been found desirable to increase the lengths of emitting electrode wires by adding a series of wires 74 which are supported on the main emitting wires 12 and which are parallel to each other and at a distance approximately equal to the distance of the emitting wires from the collecting electrode. The outermost wire is positioned inwardly about one-half the distance between the parallel wires (i.e., from 1 to 3 inches) from the outer

3,130,945

5

frame member on the collecting grid to take full advantage of all the ionized particles which are produced. The radially directed emitting wires 12 are used to electrically connect the non-intersecting wires 74 together. However, the emitting wires 12 should be fewer and much less closely spaced than collecting grid wires 14 in order to avoid electrical symmetry. If the configuration of the emitting electrode wires 12 and the collecting electrode wires 14 are identical, no lifting force is provided.

A further embodiment is shown in FIGURE 4 (Sheet 2) which is identical with the form shown in FIGURE 3 except that the structure is equipped with positive dihedral for greater stability. Center section 60 is used as a center load carrying or cabin section and the rigid support sections are attached thereto so as to tilt upwardly to form a small angle α. Collecting grids 14 and their associated emitting electrodes 12 on opposite sides of center section 60 are thus angularly related.

This particular craft, because of its horizontal symmetry, is well adapted to be equipped with dihedral in the fore and aft direction as well as in the lateral direction. FIGURE 5 (Sheet 4) represents in an exaggerated schematic form an apparatus of this type. In FIGURE 5, the central section as shown in FIGURE 4 has been omitted and the four collecting grids 14 are of a triangular shape with the inverted apex or nadir 69 of the system at the bottom and center of the apparatus. Separate emitting wires 12 are mounted from masts 71 supported centrally of the side edges and at the nadir. Each of the four collecting grids may be insulated from each other by a gap or insulating material and variable resistance incorporated in their lead-in connections (not shown) to the power supply. By independently varying the resistance of the collecting grids the craft, which is here assumed rigid, may be tilted in any direction.

FIGURES 6 and 7 (Sheet 3) are top plan and side elevation views respectively of a further embodiment which has a negative dihedral. In this embodiment, the collecting grid frame comprises four outer peripheral lightweight wooden or metal members 600, 601, 602 and 603 which are mounted in a lower plane and four inner members 604, 605, 606 and 607 which are parallel to the respective outer members but in a plane higher than the plane containing the outer members. In an embodiment where the outer peripheral members were three feet long, the vertical distance between the planes carrying the inner and outer members was four inches. This negative dihedral has been found to provide greater stability during hovering flight than the positive dihedral though the positive dihedral appears to provide equally good stability for horizontal flight.

The collecting grid is divided into four equal areas by diagonally oriented frame assemblies 608, 609, 610 and 611. The collecting grid area visible in FIGURE 7 is bounded by rigid frame members 612 and 613 of diagonal frame assemblies 608 and 609 respectively and inner and outer frame members 604 and 600. The collecting grid, as pointed out above, may be a crossed grid of wires. The other three collecting grid areas are of identical size and construction.

Inside of inner frame members 604, 605, 606 and 607, no collecting grid screen is provided and the space may be left open or if desired, covered with a lightweight foil of insulating or conducting material. This air-tight foil forms a pocket under which a pressure appears to build up to provide added lift. The insulation material is preferred since this does not interfere with the electrical isolation between the four quadrants of the collecting grid which, as will be pointed out below, are used for guiding and/or propelling the craft.

Diagonal frame assembly 608 contains four cross braces 626, 628, 630 and 632 between frame members 612 and 613. The cross braces are made of an insulating material such as wood to thereby insulate each of the four grid sections from one another. Frame members 612 and 613

6

intersect and are secured together above and inwardly behind member 604. Members 614 and 615 also intersect and are secured together, as do members 616 and 617 and members 618 and 619. These points of intersection are joined together by four struts 620 shown in FIGURE 6. Secured to the centers of each of struts 620 is a four-sided chimney 622, each of the sides being flat sheets of a lightweight insulating material such as wood.

A center frame member 624 is mounted between the center of cross brace 626 and the top of chimney 622 along each of the diagonal frame assemblies. This construction gives adequate rigidity to prevent warpage. of the collecting grid frame assembly.

The emitting wires are illustrated diagrammatically as waving lines and make up a pattern of three parallel wires 640, 641 and 642 and one transverse wire 643 across each grid area. Four supporting masts 646, 647, 648 and 649 are mounted on cross braces 628 and secured to center frame member 624 in each of the four diagonal frame assemblies 608, 609, 610 and 611. Emitting wire 640 is supported on the upper end of each of masts 646, 647, 648 and 649 with sufficient tautness to be substantially equidistant from the collecting grid at all points.

Four supporting masts 650, 651, 652 and 653 are mounted to cross braces 630 and center frame members 624 in each of the four diagonal frame assemblies for supporting emitting wire 641. Four additional masts (not numbered) are mounted on cross braces 632 and center frame members 624 to similarly support emitting wire 642.

At the mid-points of each of the outer frame members 600, 601, 602 and 603, masts 656 are mounted to support the outer end of emitting wires 643, which extend under and in electrical contact with each of emitting wires 640, 641 and 642 to a center mast 660 which is suitably mounted to the top of chimney 622.

One electrical terminal 662 for the emitting wires is shown on the right side of the craft of FIGURES 6 and 7. Four individual electrical terminals 664, 665, 666 and 667 are provided for each of the four grid sections. If it is not desired to control the posture and movement of the craft by the four separate sections, collecting grid terminals 664, 665, 666 and 667 may all be connected together.

Also, it is obvious that the four electrically separate sections could be achieved by using four insulated emitting wire sections, either with the four separate collecting grid sections or with all the collecting grid sections connected together.

The foregoing craft weighed about 100 grams and with a 5 inch spacing between the emitting wires and collecting grid, was self-sustaining with a voltage of about 50 to 60 kv. and a current on the order of 2 milliamperes.

Instead of using a crossed wire mesh construction for the collecting grid as shown in detail in FIGURE 1, it has been found that tubes of conductive material having an outer diameter of about one-quarter inch are equally as good. Such tubes may be made of aluminum foil wrapped around paper or may be hollow lightweight aluminum tubing or of a similar construction. For example, material such as an air-tight nylon base fabric having an evaporated metallic coating of for example aluminum may be fabricated in the form of tubes having a wall thickness of less than 1 mil and be adapted to be inflated with air or an inert gas to form a hollow lightweight tubular member. The cross section may be circular, oval or the like; a tear drop shape as illustrated in FIGURE 8 (Sheet 3) is a preferred configuration since air flow across the tapering lower edge provides additional lift. For the craft configuration as shown in FIGURES 6 and 7, the inflated tubes of FIGURE 8 are mounted parallel to each other and to the outer and inner frame members 600 and 604, or to their corresponding members in each of the other collecting grid sections, with their ends secured to the diagonal frame assemblies 608, 609, 610 and 611.

3,130,945

7

Other emitting electrode constructions may also be used. For example, emitting wires 640–644 may have suspended from them a plurality of short wires 680 as shown in FIGURE 9 (Sheet 3) which provide a point of discharge rather than a line of discharge to thereby increase the efficiency of ionization. In FIGURE 9, only emitting wire 640 and its supporting masts 646 and 647 from the embodiment shown in FIGURES 6 and 7 are illustrated. It is to be understod that all of the emitting wires may be of similar construction to that illustrated in FIGURE 9. Each of wires 680 is about 1 to 3 or more inches in length and separated at least one inch apart. The lower ends of wires 680 are kept at a uniform distance from the collecting grid. This construction may offer some pre-ionization, though measurements show this emitting electrode construction to be about comparable to the use of plain wire as the emitting electrode.

FIGURE 10 (Sheet 4) illustrates an elevation view, and FIGURES 11 and 12 illustrate plan views of modified triangular and rectangular shaped Ionocrafts 10 respectively. The craft of FIGURE 11 is triangular in configuration and is provided with emitting wires 12 suitably supported from masts 48 as illustrated. In practice additional emitting wires may be used. Collecting grid 14 extends over a large area beneath emitting wires 12 and may be formed of crossed wires as diagrammatically illustrated.

The electromagnetic energy antenna carried by the foregoing Ionocraft embodiments may comprise a series of generally horizontal, parallel conducting elements or dipoles 70 arranged along the basic side structure on which the wires 12 and 14 of the craft are attached. Dipoles 70 may be of differing length so that the antenna provided may receive or transmit several different frequencies. For frequencies of the order of 10 megacycles, for example, several dipoles 72, 74 and 76 may be arranged as a tuned array, such as the yagi array, with one or more dipoles 72 serving as a director, dipole 74 serving as the main antenna element and dipole 76 serving as a reflector. Such antenna is highly directional and with an Ionocraft of triangular configuration, the antenna may be used with signal transmission in three separate directions simultaneously.

The antenna wires 70–76 may be small diameter rods of a conductive material such as aluminum, supported on lightweight rods or bars 78 of either a conducting or insulating material, as dictated by conventional antenna construction techniques. Additional antenna elements 80, 82 and 84 may be present as metal rods or wires separated electrically from the adjacent antenna elements by insulators 86 of a suitable light material such as wood, plastic or the like, indicated on the drawing by spaces. The various antenna elements 70–84 and insulators 86 may comprise a rigid frame forming the basic structure for the craft and inside of which the collecting grid 14 is supported and upon which the discharge electrodes 12 are mounted. The antenna elements 70–84 may be stacked vertically if desired to improve both the efficiency of the antenna and the rigidity of the basic structure. To the extent that the antenna elements may be galvanically connected together without interfering with the operation of the antenna in its conventional manner, the antenna elements are preferably connected to the same D.C. potential as collecting grid 14. Thus, the antenna elements may also augment the operation of the Ionocraft by neutralizing charged ions which provide the propelling force.

In the rectangular embodiment of FIGURE 12, the several antenna dipoles 90 have different lengths so as to be equal to one half the wave length λ of the frequency being transmitted for an entire spectrum of frequencies having different wave lengths $\lambda_1, \lambda_2, \lambda_3 \ldots \lambda_n$. Since the length of a side of the Ionocraft may be several hundred feet or greater, such construction is ideally suited for communication systems, whether operating with high or low frequencies.

8

With either of the configurations of FIGURE 11 or FIGURE 12, the view in elevation will be substantially as illustrated in FIGURE 10 where the particular antenna structure is indicated schematically and designated by reference numeral 92. A ground station antenna which is indicated diagrammatically at 94 on FIGURE 10 may be provided for directing the signals downwardly to the ground station. Antenna 94 may be of any desired conventional type and connected on the Ionocraft to the main antenna structure 92 by a suitable transmission line such as coaxial cable, twin lead lines or hollow pipe waveguide, depending upon the particular frequencies utilized. Amplifiers or frequency converters may also be provided in the transmission line where signal strength is weak. The amplifiers and/or frequency converters may be powered by well known self-contained batteries or by the power supply unit for the Ionocraft (not shown).

Referring now to FIGURES 13 and 14 (Sheet 4), a further embodiment of the invention is illustrated which has a plurality of side sections, four of which are shown curved. The contour of the curves may be parabolic or of any other shape as is conventionally used for antennas in high frequency systems such as radar or the like. In this embodiment, an outside frame of lightweight rigid members 96, 98, 100 and 102 is provided to define the contour of the antenna shape. Lightweight wires or rods 104 extend between members 95 and 102 to serve as part of the antenna. Lightweight sheet metal of a material such as aluminum may be used in lieu of wires 104 for the reflector surface if desired.

A plurality of horns 106 are illustrated in the drawings to effect simultaneous radar scanning through 360°. By oscillating the illustrated Ionocraft about its vertical center line through an angle of only 45° on each side of a center position, complete 360° scanning may be effected. Alternatively, the Ionocraft may be rotated continuously about its vertical center line and 360° scanning effected by one or more antennas. Such scanning may be effected by warped corners, reactive or propeller blasts of auxiliary power plant, or by auxiliary grids which are mounted for movement relative to the main lifting grid as will be described below. Scanning may be effected by other means as will become apparent from the following description. In lieu of or supplemental to some of the microwave antennas 106, antenna reflectors for infrared detectors may be carried on the Ionocraft. Such antennas serve to collect the infrared energy over a large area and focus such energy on a small infrared detector, and they may be of any conventional construction. The basic structure 10 between spaced antennas may contain such equipment to transmit via wireless signal channels to the ground station through ground station antenna 94, signals corresponding to the electromagnetic and radio frequency signals received.

Horizontal movement of the craft may be effected by the principles set forth in Serial No. 760,390 of Glenn E. Hagen filed September 11, 1958, by tilting the craft downwardly in the forward direction whereby the ionic propulsion force provides a horizontal force component to cause the craft to move in a horizontal direction. Tilting of the craft may easily be effected through variation of the voltage between emitting electrodes 12 and collecting grid electrode 14. For example, by electrically separating the craft into four sections of substantially equal size as illustrated in FIGURE 15 (Sheet 6), the voltage applied to two of the adjacent sections can be reduced by adding resistance in series with the current path and this will cause the lift produced by these two sections to decrease relative to the lift produced by the two other sections. Thus, horizontal movement of the craft may easily be controlled from the ground station.

For manual control of the posture and flight movement of the craft of the present invention, it has been found desirable to provide a control stick assembly which functions similar to that familiar to persons flying other types

3,130,945

9

of aircraft. The control stick must function in both the longitudinal and lateral directions simultaneously and independently. Variable control elements such as potentiometers and variable transformers (powerstats for instance) may be used for control of the present invention. The posture of the craft may be controlled by dividing the collecting grid into three or more electrically separate regions as illustrated by the embodiment shown in FIGURES 6 and 7 and by individually varying the electrical potential to each of the separate regions. The potential may be increased to act as an elevator or may be decreased to act as a spoiler, and the voltage may be increased on one side while being simultaneously decreased on the other side to increase the effectiveness of the control.

Also, the emitting wires may be divided into three or more electrically separate regions and the electrical potential individually varied to each separate region. Again the potential may be increased or decreased, and may be simultaneously increased in one region and decreased in the opposite region.

To change the voltage to an individual region of the craft, a separate power supply for each region may be provided and the variable control element for changing the output voltage may be adjusted to produce the desired voltage level. Where a single power supply is provided, variable resistances may be placed in the electrical conductors leading to the appropriate terminals on the craft. If the craft is normally airborne with resistance present in the conductors, then increased voltage can be supplied to one region of the craft by decreasing the resistance in the conductor connected to that region. A decreased voltage can be supplied similarly by increasing the amount of resistance, and combination of increased and decreased voltages may be supplied to opposite sides of the craft to increase response of the craft to the controls.

One of the more simple ways to utilize the power supplied to the craft, I prefer not to have extra resistance in the power supply circuit of the emitting wires during normal flight and to control the posture of the craft by individually adding resistance into the circuit connected to each individual region of the collector grid. Such method of control has been found to provide adequate control of the Ionocraft and a control stick assembly will be described which utilizes variable resistance elements which are conventionally known as potentiometers or rheostats.

In FIGURE 16 (Sheet 5) the collecting grid construction as shown in the preceding embodiments (see for example FIGURES 6 and 7 on Sheet 3) is illustrated with each of the four grid sections W, X, Y, and Z connected through a separate correspondingly designated potentiometer to one terminal of the power supply. The emitting wires shown diagrammatically as waving lines are connected through a throttle control potentiometer, which is used to control the maximum voltage applied to the emitting wires and all of the collecting grid sections. When this voltage exceeds a certain level but yet remains less than that which causes arcing, the craft will rise. The effect of potentiometers A, B, C and D is to controllably reduce the voltage between the emitting wires and any one or two specific grid sections to thereby reduce or subtract from the effectiveness of that portion of the craft in producing its lifting force. This then causes the craft to tilt downwardly in the direction of whichever of the grid sections has the reduced voltage.

Referring now to FIGURES 17 and 18 (Sheet 5) front and side elevations of the control stick are shown with the respective shafts of the four potentiometers labeled A, B, C and D. On each of these shafts spur gears (not shown) are provided to be driven by gear segments secured to the stick.

The control stick is mounted for pivotal movement about pin P having axis X and about pin Q having axis Y beneath, but in the same vertical plane as axis X. Pin Q is mounted with its ends in opposite side walls W of the control stick housing.

The entire stick assembly shown in FIGURES 17 and

10

18 is mounted for unitary movement in a plane perpendicular to the longitudinal axis Y of pin Q. This assembly comprises bracket F which has secured to one side face spur gear G which need have only a segment thereof with teeth to mate with the pinion gears on the shafts of potentiometers B and D. The housings for potentiometers B and D are mounted on housing walls W, and the center of the gear segment on gear G coincides with axis Y of pin Q.

The ends of pin P are mounted in opposite sides of bracket B to enable the control stick to rock in a plane perpendicular to the longitudinal axis X of pin P. The lower end of the control stick is bifurcated as shown in FIGURE 18 and adapted to pivot about pin P. Gear segment H, having its center at axis X of pin P, is secured to the control stick for driving pinions on the shafts of potentiometers A and C which are mounted on bracket F.

The foregoing construction permits the control stick to function both in a longitudinal direction and in the lateral direction simultaneously to function as an electrostatic spoiler in the sense that when the craft is airborne, the addition of resistance in the lead-in wire to a particular grid section spoils the lift of that section to thereby control the posture of the craft in flight.

In the described embodiment, stick movement was limited to about 40° by mechanical stops not shown. The pitch diameter of each gear segment G and H was about 6 inches and the pitch diameter of the pinion gears on the potentiometer shafts was about 1 inch. The potentiometer gear shafts were capable of rotating through 240°, and were spring biased to a zero resistance condition.

As is apparent from FIGURES 17 and 18, the position of the pinion gears for the four potentiometers A, B, C and D is at the exact ends of the corresponding gear segments so that when the control stick is in its illustrated vertical position, each potentiometer is rendered ineffective to add any resistance to any of the collecting grid sections. As the control stick is tilted, one of the potentiometer shafts is rotated and there is absolutely no possibility that the potentiometer to the opposite grid section can be made effective at the same time because the partial gear segment and the spring loaded potentiometer shafts are used. The length of each gear segment must be at least as large as the maximum angle through which the stick can be moved, and the pinion gears are preferably at the precise ends of the gear segments.

It was found that if the potentiometer shafts were not spring loaded, the gears would upon occasion rotate slightly so the teeth did not always mesh when the stick was moved in a direction so that the gear segment should have engaged the potentiometer pinion. By the manual control stick just described, adequate tilt of the craft is readily achieved.

The position of the craft in air may be remotely controlled from a ground station through wireless control systems which may be of any suitable known type. The horizontal position of the craft may also be controlled automatically.

For example, the position of the craft of the present invention may be automatically controlled in space through means of suitable centering or tracking apparatus operating on well known principles, such for example as are disclosed in U.S. Patent Nos. 2,513,367 to Scott or 2,604,601 to Menzel. In such tracking apparatus, one form of which is diagrammatically illustrated in FIGURE 15 (Sheet 6), a beam of electromagnetic energy, such as light or infrared, is centered on a suitable photocell 128 which generates control signals that are used to control variable impedances to reduce the voltage applied to various sections of the craft to thereby control the position of the craft in accordance with the position of the beam source at the ground station.

FIGURE 15 illustrates in detail suitable horizontal positioning control arrangement. The common grid elec-

3,130,945

11

trode 14 is connected to the negative terminal of the power supply and the emitting wires 12 are electrically separated into four sections, viz. left front LF, left rear LR, right front RF and right rear RR. Each of these sections is connected through variable impedances 130, 132, 134 and 136 respectively of the elevator control unit and the variable impedances 138 and 140 of the aileron control unit to the positive terminal of the power supply. The elevator motor 142 drives the movable contacts on variable impedances 130, 132, 134 and 136 and the aileron motor 144 controls in a similar manner values of the impedances 138 and 140. Each motor 142 and 144 may be driven by separate amplifiers 146 and 148 and pre-amp 150 in a manner as conventionally used in servo systems to position photocell unit 128 directly in alignment with a source of electromagnetic energy positioned on the ground.

Referring now to FIGURE 19 (Sheet 2), a craft having a central cabin 160 and equipped with dihedral is illustrated. The collecting grid 14 and emitting wire 12 construction may be similar to that described in connection with FIGURE 4 (Sheet 2) and be positioned on alternate sides of cabin 160. Beneath cabin 160, a suitable wheeled, skid or pontoon landing gear 162 may be provided.

Depending beneath frame members 164 and on opposite sides of cabin 160 are a pair of auxiliary grid assemblies 166 and 167 that are mounted to be operable in a generally vertical plane. Each auxiliary grid assembly 166 and 167 is provided with laterally spaced emitting wires 168 and a collecting grid within outer frame members 170 so that upon receipt of a suitable D.C. potential, a horizontal thrust is provided in the manner hereinbefore set forth.

Each auxiliary grid assembly 166 and 167 is mounted on frame members 164 for independent rotational movement about substantially horizontal axes 172 and 173. With the emitting wires 168 of both auxiliary grid assemblies facing in the same direction, the craft will proceed in the direction toward the emitting wires. With the emitting wires 168 of auxiliary grid assembly 167 facing in a rearward direction and emitting wires of grid assembly 166 facing in a forward direction as illustrated in FIGURE 19, the craft will revolve about an axis mid-way between the effective centers of the two grid assemblies. If the craft is simultaneously tilted in a cyclical manner, an effective radar antenna searching motion is provided which may include a large vertical angle as well as a 360° horizontal scanning operation.

Except where rotation of the craft for searching or scanning operations is a principal purpose for the craft, the emitting wires 168 of each auxiliary grid assembly 166 and 167 are mounted to face in the same direction. When landing or taking off, which is always accomplished in a vertical direction, auxiliary grid assemblies 166 and 167 are preferably pivoted into a horizontal plane. This not only retracts them to prevent interference with landing operations, but also provides a multiple deck structure to give additional lift and control of stability. Horizontal speed may be controlled by varying the angle of auxiliary grid assemblies 166 and 167 with the vertical.

As shown in FIGURE 20 (Sheet 6), the Ionocraft may comprise several decks 180, 182 and 184 each of which is of similar construction to the single-decked craft shown in FIGURES 10–14 (Sheet 4). Each of the basic structures 180, 182 and 184 may comprise different antenna types if desired. Several separate ground station antennas 186, 188 and 190 may be provided particularly where independent signals are transmitted and received by the several antennas of the Ionocraft.

In FIGURE 21 (Sheet 2), a multiple decked craft is illustrated which comprises a central cabin 200 from which two lifting grid assemblies 202 and 203 extend laterally on opposite sides which are equipped with dihedral. Above grid assemblies 202 and 203, one or more

12

pairs of similar grid assemblies 204 and 205 are supported by a suitable superstructure 208. The turning axes 212 and 213 for auxiliary grid assemblies 210 and 211 in this embodiment are substantially vertical and extend through support members 214 and 215 to the upper grid assemblies 204 and 205 to provide added rigidity to the craft structure. Retractable antennas 220 and 221 may be provided beneath cabin 200 for establishing communication channels to the ground station (not shown).

In general, it makes little difference whether the emitting wires 12 are connected to the negative or to the positive terminal of the power supply. By tests, it has been determined that with emitting wires 12 connected to the negative terminal, there is an improvement of about 5% over that obtained when the emitting wires 12 are connected to a positive terminal.

In the multiple deck constructions, it is preferable to connect emitting wires 12 and collector grid wires 14 of the adjacent decks to opposite terminals of the high voltage generator as illustrated in FIGURE 20 (Sheet 6), thus making discharge or emitting wires 12 in alternate decks positive and the collector grids negative which is the reverse of the polarity shown in FIGURE 1. In that case, tilting is effected by varying either the negative or positive potential of the corresponding emitting electrode wires and grid sections to provide a rolling movement longitudinally and laterally.

All the above mechanisms and procedures provided for manual control can be utilized for automatic control actuated by an automatic pilot director through suitable servo-mechanisms.

The tilting of the craft in the case of embodiments like those diagrammatically indicated in FIGURE 15 (Sheet 6) and 16 (Sheet 5) provides forward gliding movements much in the manner that a helicopter is propelled in a horizontal direction. Where other means are used for horizontal propulsion, such for example the auxiliary grids shown in FIGURES 19 and 21 or in conjunction with propellers or jet stream, then the tilting will be used to maintain a desirable posture in space. All these movements may be controlled automatically by conventional stabilizing and steering mechanisms borne by the craft or such movement may be accomplished from remote transmitting points either on the ground or from another airborne craft.

The maximum size of crafts of the type here involved is theoretically unlimited, except for structural considerations, since the amount of lift provided increases continuously with area. It is thus contemplated that a particularly useful function of the craft of the present invention may be to serve as means for destruction through collision oncoming vehicles and missiles through air and space. Intercontinental as well as space missiles enter the atmosphere over a target area in predictable trajectories, the terminal end of which is a substantially vertical path. Thus, the large horizontal area of the craft of the present invention is particularly suitable for the purpose of protecting sensitive target areas such as large cities, naval task forces, troop concentrations and the like by its mere physical presence during hovering operations. By maneuvering the craft laterally it is possible to protect an area much larger than the area of the craft since present detection systems give identifying information of the target area about 15 minutes prior to arrival of the missile and the lateral movement of the craft may be effected at speeds of the order of 60 miles per hour, or more depending upon the horizontal propulsion system used. If the target area is vast, several Ionocraft could be maintained aloft to assure collision with oncoming missiles.

While the craft may be powered through conductors extending from ground or ship towers or via microwave power transmissions, it is contemplated that lightweight power plants such as gas turbines or the like, be used to drive suitable high voltage generators which are aboard the craft. As shown in FIGURE 22 (Sheet 6), turbine 230 may be so mounted that its exhaust is directed vertically

3,130,945

13

downwardly to provide additional lifting force while providing shaft rotation for producing the electrical power for the Ionocraft. Turbine 230 is here shown to be mounted for pivotal movement about the axis of shaft 232 which is driven by a tilt motor 234 to change the direction of the exhaust gases from vertical toward a horizontal direction. The entire tilt motor 234, shaft 232 and turbine 230 assembly may be mounted to be rotated in azimuth by azimuth motor 236 driving annular ring gear 238. Thus, in emergency operations where maximum horizontal speeds are desired, motors 234 and 236 may be controlled to advance the craft at higher velocities.

Other types of convention airborne power plants, such as turbine propeller combinations, may also be utilized for providing additional lift and aiding in maneuvering in the atmosphere. The turbine of FIGURE 15 may be provided with a reverse thrust device or such propellers may have a reversible pitch, and steering may be accomplished by rudders or vanes located in the jet stream of the turbine. Also, high voltage generation by radio-active isotopes is another method of obtaining the necessary high voltage energy or a primary source of ionization for the propulsion and sustenance of the Ionocraft.

It is also contemplated that this craft may be supplied with electrical power transmitted to the Ionocraft while in flight by microwaves. It has been demonstrated 80% of the energy emitted from a ground station microwave antenna array can be collected in the form of heat by airborne vehicles. In this case, such heat may be readily converted into high voltage by conventional means such as turbines operating high voltage generators, suitable thermocouples and vibrator-transformer convertors or the like. The use of high power microwave amplifiers, such as Amplitrons (Raytheon Co.), for power transmission via microwaves can provide the requisite power for a craft of this type. Therefore, it may be not essential that a self contained power unit be carried by the craft for special uses.

In the preferred form of the craft adapted for military purposes, directional detecting apparatus such as radar or infrared equipment will be carried by the craft to enable an antenna on the craft to lock-on any target object in air and space for the purpose of guiding the craft into the path of such oncoming target object. An Ionocraft of sufficient lift capacity may carry its own computers to process the electromagnetic information to provide the necessary impulses to the controls of the propulsive means to place the craft in the path of collision. Such craft may also be guided from the surface of the earth or from an airborne vehicle in flight, by remote control means to accomplish the collision with an oncoming object.

Explosives may be carried by the Ionocraft for destroying such oncoming objects if the mass of the Ionocraft is inadequate for destructive purposes. Such explosives may be of any known type and adapted to be detonated either upon impact or by proximity fuses where desired. Other types of countermeasures or defensive devices for causing premature explosions of the warhead of a missile may be carried by the Ionocraft as occasions arise.

The invention may be embodied in other specific forms without departing from the spirit or essential characteristics thereof. The present embodiments are therefore to be considered in all respects as illustrative and not restrictive, the scope of the invention being indicated by the appended claims rather than by the foregoing description, and all changes which come within the meaning and range of equivalency of the claims are therefore intended to be embraced therein.

What is claimed and desired to be secured by United States Letters Patent is:

1. In combination with flying apparatus composed of a structure supporting discharge electrodes for causing adjacent air molecules to become electrically charged spaced from a grid of electrical conducting means for neutralizing the charge on electrically charged molecules whose charge was caused by said discharge electrode and

14

means for applying a high D.C. potential between said electrodes and said grid to cause air to move from said electrodes toward said grids to provide a propulsion force for said flying apparatus, and antenna means for use with radio frequency energy signals, said antenna comprising a plurality of elements which serve also as part of said structure for the flying apparatus by being connected electrically to one terminal of said high D.C. potential to assist in providing propulsion force for said apparatus.

2. In combination with flying apparatus composed of a structure supporting discharge electrodes for causing adjacent air molecules to become electrically charged spaced from a grid of electrical conducting means for neutralizing the charge on electrically charged molecules whose charge was caused by said discharge electrode and means for applying a high D.C. potential between said electrodes and said grid to provide a propulsion force for said flying apparatus, an antenna for use with radio frequency energy signals, said antenna comprising a plurality of elements which serve also as part of said structure for the flying apparatus by being connected electrically to one terminal of said high D.C. potential to assist in providing propulsion force for said apparatus, and the flying apparatus structure having a configuration such that at least one side has a length greater than one-half the wave length of the radio frequency energy transmitted by the antenna.

3. The combination as defined in claim 2 wherein said antenna comprises a plurality of conductive elements having differing lengths whereby said antenna is tuned to different frequencies of radio frequency energy.

4. The combination as defined in claim 2 wherein said antenna comprises elements which serve as part of the mechanical structure of the neutralizing grid.

5. In combination with flying apparatus composed of a structure supporting discharge electrodes for causing adjacent air molecules to become electrically charged spaced from a grid of electrical conducting means for neutralizing the charge on electrically charged molecules whose charge was caused by said discharge electrode and means for applying a high D.C. potential between said electrodes and said grid to provide a propulsion force for said flying apparatus, said flying apparatus having on a plurality of different sides thereof antenna means for directional transmission of radio frequency signal energy, the length of a side being greater than one-half the wave length of the signal energy being transmitted, and means for effecting a scanning movement of said antenna means by rotating said flying apparatus about a vertical axis.

6. In combination: flying apparatus comprising a structure supporting discharge electrodes for causing adjacent air molecules to become electrically charged spaced from a grid of electrical conducting means for neutralizing the charge on electrically charged molecules whose charge was caused by said discharge electrode and means for applying a high D.C. potential between said electrodes and said grid to provide a propulsion force for said flying apparatus; said apparatus being divided electrically into different sections with separate circuits individual to each of said sections for connecting said D.C. potential to the electrodes of the respective section; means including highly directional radiant energy sensitive means on said flying apparatus for receiving energy from an electromagnetic energy source at a control station, mounted on said apparatus for generating control voltages; and circuit means for connecting said control voltages to impedance varying means to reduce the voltage to various ones of the sections of said apparatus to cause said apparatus to assume a position determined by the energy source at said control station.

7. A system and apparatus for effecting destruction of flying vehicles including a flying apparatus composed of a large area structure supporting discharge electrodes for causing adjacent air molecules to become electrically charged spaced from a grid of electrical conducting means which neutralize the charge on electrically charged molecules whose charge was caused by said discharge electrode

15

and means for applying a high D.C. potential between said electrodes and said grid to provide a propulsion force for said flying apparatus; tracking means carried by said flying apparatus for generating control signals, means responsive to said control signals for guiding said flying apparatus into the path of said flying vehicles and to effect destruction thereof by impact.

8. In flying apparatus composed of a structure supporting discharge electrodes for causing adjacent air molecules to become electrically charged spaced from a grid of electrical conducting means for neutralizing the charge on electrically charged molecules whose charge was caused by said discharge electrode and means for applying a high D.C. potential between said electrodes and said grid to cause air to move from said electrodes toward said grid to provide a propulsion force for said flying apparatus, structural supporting means for said neutralizing grid separating said grid into two substantially equal sized grid areas, and means securing said grid areas in an angularly related position to equip said apparatus with dihedral for stabilization.

9. In flying apparatus composed of a structure supporting discharge electrodes for causing adjacent air molecules to become electrically charged spaced from a grid of electrical conducting means for neutralizing the charge on electrically charged molecules whose charge was caused by said discharge electrode and means for applying a high D.C. potential between said electrodes and said grid to cause air to move from said electrodes toward said grid to provide a propulsion force for said flying apparatus, structural supporting means for said neutralizing grid separating said grid into two pairs of grid areas, the grid areas of each pair being of substantially equal size and on opposite sides of the apparatus, and means securing the grid areas of each pair in an angularly related position to equip said apparatus with dihedral in both longitudinal and lateral directions.

10. The apparatus as defined in claim 9 wherein each grid area is planar and the center of the craft is higher than the outer peripheral edges to thereby provide a negative dihedral.

11. The apparatus as defined in claim 9 wherein each grid area is planar and the center of the craft is lower than the outer peripheral edges to thereby provide a positive dihedral.

12. In flying apparatus composed of a structure supporting discharge electrodes for causing adjacent air molecules to become electrically charged spaced from a grid of electrical conducting means for neutralizing the charge on electrically charged molecules whose charge was caused by said discharge electrode and means for applying a high D.C. potential between said electrodes and said grid to cause air to move from said electrodes toward said grid to provide a propulsion force for said flying apparatus, the improvement wherein the collecting grid comprises a plurality of lightweight thin walled tubes having an outer surface of conductive material, said tubes being inflated and spaced to provide neutralization of the charged particles and permit air flow between adjacent tubes.

13. In flying apparatus composed of a structure supporting discharge electrodes for causing adjacent air molecules to become electrically charged spaced from a grid of electrical conducting means for neutralizing the charge on electrically charged molecules whose charge was caused by said discharge electrode and means for applying a high D.C. potential between said electrodes and said grid to cause air to move from said electrodes toward said grid to provide a propulsion force for said flying apparatus, the improvement wherein the discharge electrodes comprise wires of conductive material mounted above the neutralizing grid, and a plurality of short wires mounted on the discharge electrodes with free ends thereof suspended beneath the discharge electrodes at substantially equal distances from said neutralizing grid to thereby provide point sources for causing ionization.

16

14. In flying apparatus composed of a structure supporting discharge electrodes for causing adjacent air molecules to become electrically charged spaced from a grid of electrical conducting means for neutralizing the charge on electrically charged molecules whose charge was caused by said discharge electrode and means for applying a high D.C. potential between said electrodes and said grid to cause air to move from said electrodes toward said grid to provide a propulsion force for said flying apparatus, an auxiliary discharge electrode and neutralizing grid structure supported by said flying apparatus for movement relative to said flying apparatus about a pivot axis for controlling the posture and direction of movement of said apparatus.

15. In flying apparatus composed of a structure supporting discharge electrodes for causing adjacent air molecules to become electrically charged spaced from a grid of electrical conducting means for neutralizing the charge on electrically charged molecules whose charge was caused by said discharge electrode and means for applying a high D.C. potential between said electrodes and said grid to cause air to move from said electrodes toward said grid to provide a propulsion force for said flying apparatus, a pair of auxiliary discharge electrode and neutralizing grid structures supported at spaced positions on said flying apparatus, each of said grid structures being mounted for movement about pivot axes for controlling the posture and direction of movement of said apparatus.

16. Apparatus as defined in claim 15 wherein said pivot axes are in a horizontal plane.

17. Apparatus as defined in claim 15 wherein said pivot axes are vertical.

18. In flying apparatus composed of a structure supporting discharge electrodes for causing adjacent air molecules to become electrically charged spaced from a grid of electrical conducting means for neutralizing the charge on electrically charged molecules whose charge was caused by said discharge electrode and means for applying a high D.C. potential between said electrodes and said grid to cause air to move from said electrodes toward said grid to provide a propulsion force for said flying apparatus, a rigid frame for said neutralizing grid including intersecting support members lying in a vertical plane, and air foil means secured to said intersecting support members to provide added stabilization to said apparatus.

19. In flying apparatus composed of a structure supporting discharge electrodes for causing adjacent air molecules to become electrically charged spaced from a grid of electrical conducting means for neutralizing the charge on electrically charged molecules whose charge was caused by said discharge electrode and means for applying a high D.C. potential between said electrodes and said grid to cause air to move from said electrodes toward said grid to provide a propulsion force for said flying apparatus, a polygonal rigid inner and outer frame for said neutralizing grid, a center portion of said grid surrounded by said inner frame being without the electrical conducting means forming the grid, and diagonally disposed frame assemblies extending between corners of said outer frame and said inner frame.

20. Apparatus as defined in claim 19 wherein the apparatus is provided with negative dihedral and the center portion is covered with an air foil to provide a pressure wave beneath the craft for added lift.

21. In combination a base member; bracket means supported on said base member for movement about a first pivot axis; a first gear segment on said bracket; a first pair of electrical circuit elements having parameters which are variable in magnitude by rotation of a shaft individual to each circuit element; a gear on each of said shafts; said gears being positioned at opposite ends of the gear segment on said bracket so that only one shaft is rotated from a reference position at any moment; a control stick mounted for pivotal movement on said bracket about a

3,130,945

17

second pivot axis perpendicular to said first pivot axis; a second gear segment on said control stick; a second pair of electrical circuit elements similar to said first pair of electrical circuit elements mounted on said bracket means; and separate gears for controlling said second pair of electrical circuit elements, said gears being mounted at opposite ends of said second gear segment so that only one gear is rotated from a reference position at any moment.

22. The combination as defined in claim 21 wherein each electrical circuit element contains a shaft spring loaded to a reference position.

23. The combination as defined in claim 22 wherein the electrical circuit elements are variable resistances.

24. The combination as defined in claim 23 together with flying apparatus composed of a structure supporting discharge electrodes for causing adjacent air molecules to become electrically charged spaced from a grid of electrical conducting means for neutralizing the charge on electrically charged molecules whose charge was caused by said discharge electrode, said grid comprising four electrically isolated sections of substantially the same size, and means for connecting one terminal of a high D.C. potential source to said electrodes and a second terminal to each of the four sections of said grid through different ones of said variable resistances.

25. In apparatus for thrust generation in atmosphere, composed of spaced emitting and collecting electrodes energized with a voltage sufficiently high to be capable of causing ionization for effecting an ionic discharge from an emitting electrode to produce thrust on at least one of the electrodes of said apparatus arising from elastic molecular and particle collisions occurring in the space between said electrodes during molecule and particle movement in the direction from the emitting electrode to the collecting electrode, said collecting electrode being composed of crossed grid wires of conductive material forming an open surface that is generally normal to the direction of molecule and particle movement, the improvement wherein said crossed grid wires have a diameter of at least 8 mils.

26. The apparatus as defined in claim 25 wherein the collecting electrode ends are covered by a layer of conductive material.

27. In flying apparatus composed of a structure supporting discharge electrodes for causing adjacent air molecules to become electrically charged spaced from a grid of electrical conducting means for neutralizing the charge on electrically charged molecules whose charge was caused by said discharge electrode; and means for applying a high D.C. potential between said electrodes and said grid to cause air to move from said electrodes toward said grid to provide a propulsion force for said flying apparatus; the improvement wherein said neutralizing grid comprises outer peripheral frame members and a centrally positioned inner frame between which the conducting means comprising the neutralizing grid are mount-

18

ed. and which are positioned in different horizontal planes to thereby provide the craft with dihedral; and said discharge electrodes comprise a plurality of substantially parallel wires, each mounted parallel to the outer peripheral frame members and in different horizontal planes so that all wires are substantially equidistant from said neutralizing grid.

28. A system and apparatus for effecting destruction of flying vehicles including a flying apparatus composed of a large area structure supporting discharge electrodes for causing adjacent air molecules to become electrically charged spaced from a grid of electrical conducting means which neutralize the charge on electrically charged molecules whose charge was caused by said discharge electrode and means for applying a high D.C. potential between said electrodes and said grid to provide a propulsion force for said flying apparatus; auxiliary discharge electrode and neutralizing grid means mounted for pivotal movement for controlling the forward direction of movement; tracking means carried by said flying apparatus for generating control signals, means responsive to said control signals for guiding said flying apparatus into the path of said flying vehicles and to effect destruction thereof by impact.

29. A system and apparatus for effecting destruction of flying vehicles including a flying apparatus composed of a large area structure supporting discharge electrodes for causing adjacent air molecules to become electrically charged spaced from a grid of electrical conducting means which neutralize the charge on electrically charged molecules whose charge was caused by said discharge electrode and means for applying a high D.C. potential between said electrodes and said grid to provide a propulsion force for said flying apparatus; a gas turbine mounted on said apparatus with its exhaust gases directed downwardly during hovering operation and means for deflecting said exhaust gases in a direction providing a horizontal propulsion force to increase to the forward velocity of said apparatus; tracking means carried by said flying apparatus for generating control signals, means responsive to said control signals for guiding said flying apparatus into the path of said flying vehicles and to effect destruction thereof by impact.

References Cited in the file of this patent

UNITED STATES PATENTS

2,495,748	Matson	Jan. 31, 1950
2,503,109	Harris	Apr. 4, 1950
2,598,064	Lindenblad	May 27, 1952
2,613,887	Woods	Oct. 14, 1952
2,842,645	Dalgleish et al.	July 8, 1958
2,888,189	Herb	May 26, 1959
2,892,949	Hardy	June 30, 1959
2,949,550	Brown	Aug. 16, 1960

FOREIGN PATENTS

1,174,334	France	Nov. 3, 1958

13
News Articles

Inventors flock to exposition

CR: R. Reid

DAILY NEWS, Ft. Walton Beach, FL - May 4, 1990

MONROEVILLE, Pa. (AP) — If necessity is the mother of invention, is silliness the father?

One might think that, based on some of the gizmos and gadgets on display Thursday when a convention to market inventions got under way.

Among the items: a portable fire escape that can be anchored inside a window in a burning high-rise, a trap designed to lure fleas off the carpet and a device to repel mosquitoes with sound waves.

Also displayed: a plastic gun that clears clogged drains with a blast of compressed air, a sand-laden trash can that won't tip over, and golf shoes with playing tips stitched across the toes. Would you believe a zipper tie?

Promoters and inventors from around the world showed off their creations at the Monroeville Expo Mart in suburban Pittsburgh, seeking investors and marketers.

The start of the three-day exhibition marked the unveiling of the engine created by Yoshiro Nakamatsu, who became known as the "Edison of Japan" after inventing the computer floppy disk and the digital watch.

Nakamatsu said his pollution-free engine, called Enerex, runs on tap water alone and can create three times as much power as a standard gasoline engine.

"It will generate electricity for any purpose," he said. "Petroleum will exhaust in 100

Yoshiro Nakamatsu holds his Enerex, a device he claims will enable automobiles to run on plain tap water instead of gasoline, in Monroeville, Pa., on Thursday.

years."

He said he plans to modify the engine and build a special vehicle powered by it.

Nakamatsu displayed some of his other inventions, including packages of "brain food," which he said contains "good elements" to encourage clear thinking. The snack food, which tastes like seaweed, is sold in Japan.

Nakamatsu, 62, said listening to Beethoven's Fifth Symphony and swimming have helped him maintain the creativity that has enabled him to obtain more than 2,000 patents in the United States and Japan.

"My best place to create inventions is under the water," he said. "So I swim every day."

Miracle No-Fuel Electric Engine Can Save U.S. Public $35 Billion a Year in Gasoline Bills

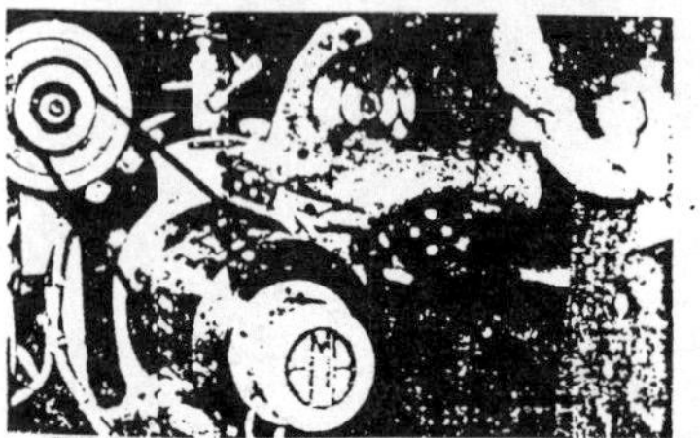

THIS IS THE 'EMA' which can run perpetually on batteries that recharge themselves, develop 1,000 horsepower per

BY TOM VALENTINE
(Second of Two Articles)

An inventor and his small but stubborn team of engineers have devised the most revolutionary technological advance in the history of mankind: A power source that uses no fuel.

As reported exclusively last week in TATTLER, the astonishing new system creates electricity without consuming the world's dwindling supply of fossil fuel, without creating pollution and without using costly and unsightly transmission lines.

The first and most vital outcome of the theory that is forcing the science of physics to revise its fundamental assumptions is the "EMA" electric engine -- a power plant that dooms noisy, dirty gasoline motors.

That means that the nation would no longer have to consume expensive and befouling gasoline. According to figures from the American Petroleum Institute, the anticipated consumption this year is 100 billion gallons, at least $35 billion worth at the pump.

Invented by Edwin Gray Sr., 48, of the engine has been tested and is being perfected by him and his associates in EvGray Enterpris-

The silent, pollution-free EMA recycles its own energy and can run indefinitely.

Gray's prototype is powered by four six-volt batteries which "will wear out before they'll run down," as the inventor puts it.

"WE CAN GO up to 1,000 horsepower with a single unit, or down to a miniature toy size."

The latter units, in fact, will be among the first products EvGray manufactures. They'll run off a tiny battery unit.

How?

Gray and his engineers, Richard Hackenberger and Fritz Lens, explained to TATTLER that they have found a way to use both the positive and negative particles of electricity separately.

The technicians demonstrated for this reporter the phenomenon of electromagnetic repulsion -- the power source for the EMA.

TWO MAGNETS, each weighing 1 3/4 pounds were repelled apart with an explosive force, but the magnets did not heat up and 99 per cent of the energy recycled to the battery.

Hackenberger, an electronics specialist, explained: "A series of high-voltage energy 'spikes' are developed by our circuitry. These energy units are transferred to a control unit, which acts much like a distributor in an internal combustion engine."

The control unit is the key to the motor's efficiency. It regulates the energy spikes to determine the polarity (north or south) and directs the voltage into selected electromagnets in the main unit.

"Every time a magnet is charged, most of the energy is recycled back into the batteries without losing power," Hackenberger said.

THE EMA has been tested thoroughly. Its efficiency is undisputed.

"Engineers and physicists who see it operate have a hard time believing their eyes," Gray said. "One professor from UCLA insisted we had some sort of laser beam running it, and even though we moved it from room to room, he wouldn't believe it."

While the motor was running, Gray spun it around in a complete circle to demonstrate that it operated at any position.

The experimental model engine is 42 inches long, 18 inches wide and 22 inches high about the size of a standard six-cylinder motor.

IT TURNED better than 2,500 revolutions per minute for more than 20 minutes. The power input came from the four six-volt batteries. At the end of the trial they were tested and found to be as fully charged as they were at the beginning.

It generated 100 horsepower and 66 pounds of constant torque. The brake horsepower tests out at 32.06.

The motor has only two bearings which require lubrication, so maintenance costs will be minimal; it operates at a maximum temperature of about 170 degrees and is cooled by compressed air.

It started with the flick of a switch. It can be accelerated or allowed by any mechanical device which programs the control unit. This means the customary foot pedal could be used for driving purpose-

By TOM VALENTINE
Copyright 1972 The National Tattler
(First of a series)

A CALIFORNIA INVENTOR has found a way to create limitless electric power without using up fuel -- potentially the greatest discovery in the history of mankind.

Edwin Gray Sr., 48, has fashioned working devices that could:

•Power every auto, train, truck, boat and plane that moves in this land -- perpetually.

Warm, cool and service every American home -- without erecting a single transmission line.

•Feed limitless energy into the nation's mighty industrial system - forever.

•And do it all without creating a single iota of pollution.

Already, the jovial, self-educated Gray is forcing scientists to uproot their most cherished beliefs about the nature of electromagnetism.

Eventually, his discovery will transform the economic base upon which the society of the entire planet has rested up to this point.

Despite the ever-present danger from the petroleum and other power giants who face business extinction within the decade because of his invention. Gray and his associates in EvGray Enterprises have demonstrated its worth publicly -- an act requiring great courage.

And TATTLER is proud to report for the first time in America the complete nature of Gray's astounding system.

Displaying the kind of open honesty that made America great, Gray and his partners stess the fact that they want the whole world to benefit from their new technology.

"I WON'T ALLOW it to be bought up and buried by big money interests," Gray told TATTLER during the exclusive demonstration.

"I tried for 10 years to get American interests to pay some attention, but I've been tossed out of more places than most people ever think of going into."

Neither government agencies nor private enterprise would listen to Gray, so he turned in frustration to foreign interests. The innovative Japanese were eager to listen.

"AS SOON AS word got out that the Japanese were interested in what we're doing, the Americans started flocking around."

Today, the small shop facility in ____ is crawling with visitors from every segment of U.S. industry and finance.

"The big money boys from Wall Street started coming around," Gray said, with a touch of defiance in his tone.

"A bunch of them came in and suggested I file bankruptcy and get rid of all my backers and friends. Then they talked about giving me 20 million shares of a new corporation at $25 a share."

Gray was being offered a deal worth more than $4 billion -- on paper.

"THAT SURE sounded rich, but I know darn well they would have fixed it up to sell that corporation off somewhere for a dollar and leave me holding 20 million shares of nothing."

Edwin Gray Sr.

The key men at EvGray include Richard B. Hackenberger, as electronics engineer who formerly worked for Sony and Sylvania corporations and the U.S. Navy; and Fritz Lens, a former Volkswagen mechanic who knows nearly as much about the fantastic electrical system as Gray.

All the corporate officers agreed that they are determined to get around the money roadblocks and bestow the discovery upon the world.

TATTLER was given a thorough demonstration of Gray's "impossible but-true" methods for using electricity.

THE FIRST demonstration proved that Gray uses a totally different form of electrical current -- a powerful, but "cold" form of the energy.

A six-volt car battery rested on a table. Lead wires ran from the battery to a series of capacitors which are the key to Gray's discovery. The complete system was wired to two electro-magnets, each weighing a pound and a quarter.

"Now, if you tried to charge those two magnets with juice from that battery and make them do what I'm going to make them do, you would drain the battery in 30 minutes and the magnets would get extremely hot," Gray explained. "I want you to watch what happens."

As Lens activated the battery, a voltmeter gradually rose to 3,000 volts. At that point, Gray closed a switch and there was a loud popping sound. The top magnet hurled into the air with tremendous force and was caught by Hackenberger. A terrific jolt of electricity had propelled the top magnet more than two feet into the air -- but the magnet remained cold.

"The amazing thing," Hackenberger said, "is that only 1 per cent of the energy was used -- 99 per cent went back into the battery.

GRAY EXPLAINED, "The battery can last for a long time, because most of the energy returns to it. The secret to this is in the capacitors and in being able to split the positive."

When Gray said "split the positive" the faces of two knowledgeable physicists screwed up in bewilderment.

(Normally, electricity consists of positive and negative particles. But Gray's system is capable of using one or the other separately and effectively.)

"He means we have to rewrite the physics textbooks," Hackenberger grinned. It has been the engineer's job in recent month to formulate Gray's system and put it in writing.

"That's not an easy task because this system actually defies everything I've ever learned."

Gray said, "I never had no schooling in electronics or physics, so nobody told me it was impossible."

THE "IMPOSSIBLE" part of the demonstration was the lack of heat generated in the magnet. Heat is one of the biggest problems faced by electrical technology. Also "Impossible" is the fact that only the "positive" nature of the energy was used.

"This thing is in its infancy," Gray explained. "When the full potential of American techbology starts working with it, the results will astound everyone."

A further proof that he has an unusual source of power with unlimited potential was demonstrated next.

"We've been popping those magnets apart for the past 18 months with that same battery and it's still got a full charge. Now I want you to watch this."

Gray showed this TATTLER reporter a small 15-amp motorcycle battery. It was hooked up to a pair of his capacitors which in turn were hooked up to a panel of outlets.

HE FLICKED a switch and the tiny battery sent a charge into the capacitors. He then plugged in six 15-watt electric light bulbs on individual cords -- a 110-volt portable television set and two radios. The bulbs burned brightly, the television played and both radios blared -- and yet, the small battery was not discharging.

"You couldn't begin to get all this current out of that battery under ordinary circumstances," Gray said.

"This is the most amazing thing I've ever seen," exclaimed C.V. Wood Jr., president of the McCulloch Oil Corporation, who was also present at the demonstration. He began looking around for hidden outlets from the wall.

"MAY I PROVE it doesn't come from any wall plug?" Gray offered.

A 40-watt light bulb screwed into an ordinary extension socket was plugged into the panel powered by Gray's system. The bulb lit, then Gray dropped it into a cylinder filled with water.

"What would be happening if this was getting ordinary power right now?" Gray asked, as he stuck his hand in the water with the glowing light bulb.

"You'd be electrocuted and that thing would be popping and sputtering until the fuses blew," Wood replied.

This reporter then put his finger into the water with the light -- no shock.

"Gentlemen, this is a new manifestation of electricity," Hackenberger said.

THE ENGINEER told the astounded onlookers that no laws of physics were being violated, but a new application of electricity has been discovered and put to work.

Gray, one of 14 children, comes from Washington, D.C. As a small boy, he was fascinated by electricity, magnets and gadgets in general.

"I really got excited about electricity when they tested the first radar across the Potomac in 1954. I was 11 years old then and visions of Buck Rogers danced in my head."

He learned about radar during his World War II hitch in the Navy and "I've been messing around with coils and capacitors ever since."

HE LEARNED to "split the positive" in 1958 and spent the next dozen years finding ways to put his discovery to work.

Any abbreviated explanation of Gray's system is an oversimplification of the technical aspects of this tremendous breakthrough, but some of the best minds in the U.S. are now working with Gray to further improve his discovery.

Gray held the 40-watt bulb up out of the water and said: "You know to light this bulb takes millions of dollars in power plant facilities, transmission lines and circuitry. With my capacitors, I can provide power to any home for a couple hundred dollars."

The economic impact of that statement is beyond the imagination -- not to mention the ecology and anti-pollution benefits.

NEXT WEEK: Electro-magnetic automobile engine.

THE NATIONAL TATTLER
Page 5 July 1, 1973

Modified Kure-Tekko Unit, With Top Magnetic Spinner

The Japanese Kure Tekko permanent magnet motor is based on utilizing a uniform spiral magnetic stator which forces a magnetic wheel or rotor to revolve from a high repulsion potential to a lower magnetic repulsion potential, as indicated in the photos below.

From a minimum entrance (starting) gap of about 1/4", the rotor magnets seek to revolve the rotor from the high repulsion zone to a far lower repulsion zone at the 1-1/4" exit gap, as indicated in the photos.

The original Kure Tekko unit called for an electromagnet at the top of the unit to force the (single) rotor magnet into the small air gap, but there are operational problems involved in this method. The iron core of the electromagnet is attracted to the rotor magnet, so that the electromagnet must produce a higher-than-normal repulsion force to overcome this magnet-to-iron attraction.

In this present design, an attraction spinner, at the top of the unit revolves independently of the main rotor to attract each of the rotor magnets and drives them into small air gap to start each rotational cycle.

The attraction magnetic spinner is revolved by a small 12 volt D.C. motor, which is powered by a 12 volt NiCad battery pack, and revolves at its own natural speed rate, as indicated in the photos below.

The Kure Tekko design presents an attractive configuration which offers several possibilities for operational improvements, including multi-function and over-unity output performance.

To the best of our knowledge, this present magnetic motor prototype is the **first operational Kure Tekko type unit, outside of Japan.**

E-Z Easy Rotor Loading Arrangement with (Hinged Trunnion-Half)

Although this prototype unit does operate continuously at about 60 rpm for the main rotor, it cannot be considered as a totally successful application of the K-T magnetic stator spiral, since the major portion of the rotor's torque is provided by the top attraction spinner which "pulls" the rotor magnets successively into the small air gap to start each of the rotor's rotational cycles.

While not being a fully successful application of the original Kure

Tekko magnetic motor concept, the smooth and continuous rotation of the rotor does indicate that such an arrangement can be considerably improved, especially when the present ceramic, Ba-Fe permanent magnets are replaced with NIB (neodymium-iron-boron) permanent magnets.

Another design feature in this present prototype is the successively increasing spacing (gap) between each of the permanent magnets in the K-T spiral which reduces the natural coercivity factor between each of the adjacent magnets. This uniformly increasing space between each of the K-T spiral magnets also contributes to the magnet repulsion differential which the rotor magnets are exposed to.

The original concept behind the Kure Tekko magnetic motor, sometimes referred to as the "Magnetic Wankel" is well founded since it was based on attempting to improve the drive motor to make electric cars successful rather than to wait for improvements in electric batteries. Improvements in electric batteries have become somewhat like the weather, battery engineers talk a lot about it, but nothing ever changes, probably due to the negative economics of such improvements.

In the original Kure Tekko magnetic motor design, a single rotor samarium-cobalt magnet was utilized which was a basic design deficiency. There is no valid reason why multiple, and an equal number of rotor magnets, cannot be employed to provide rotational balance plus improved output torque for this type of unit, as in this present design prototype.

Although at a first review, the use of a free-running attraction spinner may not appear to be efficient, it should be pointed out that this **minimizes friction** to two rotational points alone, with no interconnecting belt or gear drives between the two rotating components. Attempts have been made to use magnetic ram and crank drives to force the rotor magnets into the small air gap, without any success, so have been abandoned. These mechanical drives entailed some friction which handicapped this method.

A Pre-Tesla Tesla Coil
by Jerry L. Ziegler

My hypothesis is that in ancient times the sun was more active, spewing out many more particles in its solar wind. This caused higher electric fields on the earth which gave way on certain occasions to electrical point discharges which were especially noticeable on mountain tops. This point discharging became strong enough to produce what we call St. Elmo's fire today. These phosphorescent, dancing flames were enhanced to the point of becoming a wonder and the basis of ancient religions.

To draw down the fire of God a sacrificial pyre was made to help ionize the air, thus causing a more conductive column for the electrical charges to travel down from the skies. One might simply erect a tall pole to effect the same thing. When the prophet poured water around his altar, it helped bring the divine fire. If one lived on a flat oasis or plain, one could erect a pointed mountain, that is, a pyramid to seek the divine light.

Often the electrical fields were not strong enough to make an observable electrical flame. For instance, if a pole were erected, this might increase the conduction of charges through its top, but not enough to ionize the air so as to have an observable electric light. In order to detect this invisible current, a priest constructed a special topless box for the pole to sit in.

This box had metallic surfaces on the inside and out with a layer between made of an electrical resistive (dielectric) type material. In the *Ark of the Covenant* the two metallic plates were gold; the center layer was called *shittum* wood--essentially it was a Leyden jar. The pole naturally attracted charges which ended up on the inside layer of the capacitor, the opposite charge filling the outer layer.

One could detect the charged capacitor by touching the two metallic plates at once and receiving the shock. Two figures, cherubim, stood above the Ark box, one connected with the outer and one with the inner layer of the box, to use for detecting the voltages. These figures could be adjusted so that only a small gap existed between them. When the voltages reached high enough values there would be a spontaneous discharge of the plates, the sparking of which showed the divine light and life of the deity.

Especially at the end of the era this divine life was weakening and needed doctoring so that the discharging would be longer, giving more divine light. To facilitate this the cherubims were replaced. A wire was connected to the outer layer and brought to the Ark pole and then made to circle upwards around it in a helical shape.

At its end a narrow gap was put between the wire and the Ark pole for the discharge. To make this the more terrifying the wire was shaped as a snake. The coiled snake thus gave the whole system more inductance so that the sparking was more pronounced. In another form a snake comes from each of the two metallic surfaces, both coiling upward with a small gap between them at the top for the discharge. Both forms gave the divine life and both shapes have been used as symbols of the medical profession since early times.

printed from the September, 1927
e of Science and Invention.

Gravity Nullified

Quartz Crystals Charged by High Frequency Currents Lose Their Weight

ALTHOUGH some remarkable achievements have been made with short-wave low power transmitters, radio experts and amateurs have recently decided that short-wave transmission had reached its ultimate and that no vital improvement would be made in this line. A short time ago, however, two young European experimenters working with ultra short-waves, have made a discovery that promises to be of primary importance to the scientific world.

The discovery was made about six weeks ago in a newly established central laboratory of the Nessartsaddin-Werke in Darredein, Poland, by Dr. Kowsky and Engineer Frost. While experimenting with the constants of very short waves, carried on by means of quartz resonators, a piece of quartz which was used, suddenly showed a clearly altered appearance. It was easily seen that in the center of the crystal, especially when a constant temperature not exceeding ten degrees C. (50 degrees Fahrenheit) was maintained, milky cloudiness appeared which gradually developed to complete opacity. The experiments of Dr. Meissner, of the Telefunken Co., along similar lines, according to which quartz crystals, subjected to high frequency currents clearly showed air currents which led to the construction of a little motor based on this principle. A week of eager experimenting finally led Dr. Kowsky and Engineer Frost to the explanation of the phenomenon, and further experiments showed the unexpected possibilities for technical uses of the discovery.

Some statements must precede the explanation. It is known at least in part, that quartz and some other crystals of similar atomic nature, have the property when exposed to potential excitation in a definite direction, of stretching and contracting; and if one uses rapidly changing potentials, the crystals will change the electric waves into mechanical oscillations. This *piezo electric* effect, shown in Rochelle salt crystals by which they may be made into sound-producing devices such as loud speakers, or reversely into microphones, also shows the results in this direction. This effect was clearly explained in August, 1925 *Radio News* and December, 1919 *Electrical Experimenter*. These oscillations are extremely small, but have nevertheless their technical use in a quartz crystal wave-meter and in maintaining

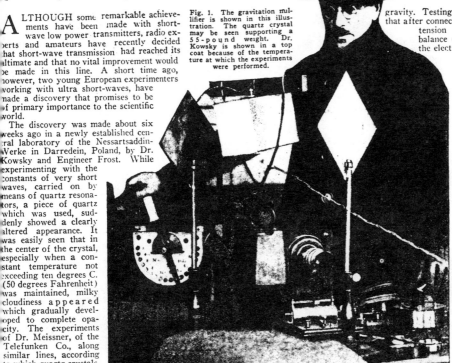

Fig. 1. The gravitation mullifier is shown in this illustration. The quartz crystal may be seen supporting a 55-pound weight. Dr. Kowsky is shown in a top coat because of the temperature at which the experiments were performed.

a constant wavelength in radio transmitters. By a special arrangement of the excitation of the crystal in various directions, it may be made to stretch or increase in length and

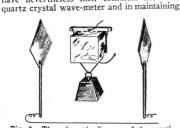

Fig. 2. The schematic diagram of the experiment is shown in this illustration. The high frequency oscillator has been omitted for clearness.

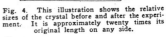

Fig. 3. This shows how the quartz crystal lost weight when subjected to the high frequency current. The original crystal was balanced on the scale.

will not return to its original size. It seems as if a dispersal of electrons from a molecule resulted, which, as it is irreversible, changes the entire structure of the crystal, so that it cannot be restored to its former condition.

The stretching out, as we may term this strange property of the crystal, explains the impairment of its transparency. At the same time a change takes place in its specific gravity. Testing it on the balance showed that after connecting the crystal to the high tension current, the arm of the balance on which the crystal with the electrical connections rests, rose into the air. The illustration, Fig. 3, shows this experiment.

This pointed the way for further investigation and the determination of how far the reduction of the specific gravity could be carried out. By the use of greater power, finally to the extent of several kilowatts and longer exposure to the action, it was found eventually that from a little crystal, 5 by 2 by 1.5 millimeters, a non-transparent white body measuring about ten centimeters on the side resulted, or increased about 20 times in length on any side (see Fig. 4.) The transformed crystal was so light that it carried the whole apparatus with itself upwards, along with the weight of twenty-five kilograms (55 lbs.) suspended from it and floating free in the air. On exact measurement and calculation, which on account of the excellent apparatus in the Darredein laboratory could be readily carried out, it was found that the specific gravity was reduced to a greater amount than the change in volume would indicate. Its weight had become practically negative.

There can be no doubt that a beginning has been made toward overcoming gravitation. It is to be noted, however, that the law of conservation of energy is absolutely unchanged. The energy employed in treating the crystal, appears as the counter effect of gravitation. Thus the riddle of gravitation is not fully solved as yet, and the progress of experiments will be followed further. It is, however, the first time that experimentation with gravitation, which hitherto has been beyond the pale of all such research, has become possible, and it seems as if there were a way discovered at last to explain the inter-relations of gravity with electric and magnetic forces, which connection, long sought for, has never been demonstrated. This report appears in a reliable German journal, "Radio Umschau."

Fig. 4. This illustration shows the relative sizes of the crystal before and after the experiment. It is approximately twenty times its original length on any side.

Don't fail to see our next issue regarding this marvelous invention.

14
Free-Energy Devices and Pop Culture

FREE ENERGY DEVICES

AN AUSTRALIAN RESEARCH PRIMER   by Peter Nielsen - Parascience Technologies

Let's get one thing straight. Nothing is for free. But EVERY-THING moves in cycles. There is no up without a down. By ingenious use of certain materials and spatial relationships, we CAN tap into one half of a full cycle. This unidirectional force = workpower. For example, a waterwheel captures one aspect of the rain cycle. If we ignore the role of evaporation and gravity, this is 'free' energy. So, what we are learning to do here is dovetail with larger systems, rather than asserting direct force which consumes local resources.

Now consider economy. Things like geothermal power are free, but expensive to contain and utilize. Ordinary electrical generators convert the motion of steam or water to voltage with far less than 100% efficiency. This is compounded by further losses in the power distribution grid. Ideally, we need something that already IS everywhere, and capable of motivating electrons (voltage) without an elaborate mechanical interface. Could you imagine a consumer tax on GRAVITY? How about the Earth's own electrostatic field.....or 'empty' space?

Our approach is radically simple: to produce a device that will of itself derive momentum from these primary existing forces. Impossible? Throw away your high school physics. Here are but a few solutions offered by previous researchers. Recent advances in superconductors will now make many of these even more commercially viable. Where are these gadgets now? Before you read any further, ask yourself what YOU would do with a free energy box. Well, some influential persons would call that a potential social crisis. Question answered.

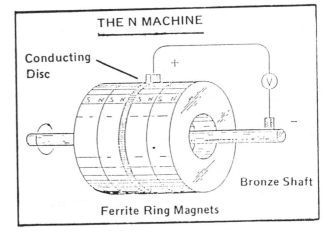

THE N MACHINE

Conducting Disc

Bronze Shaft

Ferrite Ring Magnets

HOMOPOLAR MOTOR 'N' MACHINE
Bruce De Palma

Cylindrical conducting magnet spun at high speed develops DC voltage between its periphery and axis. A fraction of this output redirected to sustain the drive motor. The rest is free. Similar to Searle effect, and Muller device in Maggie's last issue.

WATER VORTEX TURBINE
Viktor Schauberger

A counter-clockwise spiral is the path of least resistance for fluids under the pull of gravity. Conversley, by spinning the column of water in the opposing way, it actually assists the impeller by lifting itself UPWARD against the sides of the vertical tube.

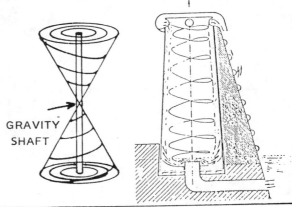

GRAVITY SHAFT

ELECTROKINETIC MOTOR
T.T.Brown

'Dielectric stress' is created with-in the atoms of an insulating mat -erial by insertion between two plates charged with high voltage DC. The entire assembly recoils on swing-arms in the direction of the positive electrode, as the atoms attempt to regain symme-try of mass by moving through space.

Atom without Applied Field

Electron Cloud

Atom with Applied Field

Displacement of Charge

FIG. 8

FIG. 7

INVENTOR
THOMAS TOWNSEND BROWN
BY

ELECTROKINETIC APPARATUS

CORONA MOTOR
Oleg Jefimenko

A 300' antenna supported by hel-ium balloons develops 2000 volts DC from altitude differential in the Earth's electrostatic field. This is switched across split field plates which alternately attract and repel an electret disk causing it to spin. Electret is a wax-like compound which has been solid-ified in a DC field to behave thereafter like a permanently charged object.

SELF-RECHARGING BATTERY
J.Bedini

Switching system concentrates high voltage spikes within the cir-cuits conductors, which are peri-odically diverted against the battery's normal flow. This in-tensified pulse acting upon the internal chemistry, liberates the bonded energies of its constit-uent ions, thereby intermittently recompensating the discharge cycle.

MOTIONAL FIELD GENERATOR
W.J.Hooper

Direct current is sent running through many layers of a single folded wire. Cancellation of opp-osing lines of force in adjacent conductors produces a secondary gravity field AND voltage grad-ient across the insulated tubular electrodes (capacitor).

Electret Rotor

Stator

Slot Effect Electret Motor

COIL

CAPACITOR

German Research: Hypothetical configuration of a secondary grav -itational field

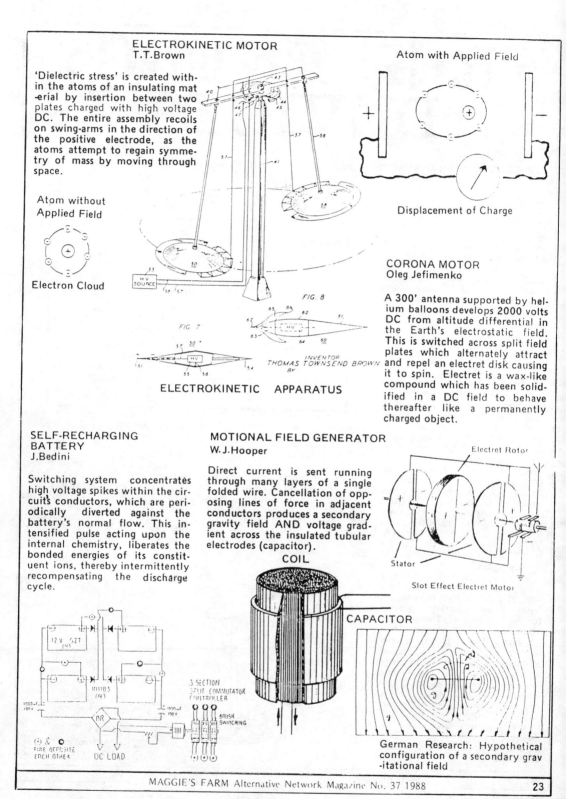

FREE ENERGY DEVICES

AND FINALLY....how to start building your own version of a free energy source, presently under evaluation right here in Australia. Warning: this abbreviated design is offered for educational purposes only. First the principle

STRESS is the mobilizing force in nature. Things and events empower their changing existence, by intermeshing their form with prevailing flows from higher to lower stress. You can feel the presence of this sustaining force in a relatively pure form, as the enlivening compressional surges which accompany thunderstorms Without it, the phenomenal world would 'run down' like a wind-up clock. It is also no accident that many of our most creative cities are located over geological faults. Historically, they are the springboards for the evolution of humanity.

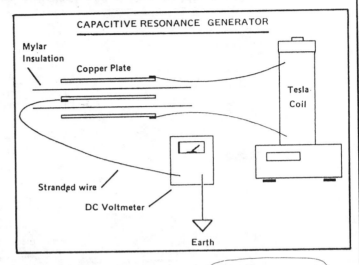

CAPACITIVE RESONANCE GENERATOR

Mylar Insulation
Copper Plate
Tesla Coil
Stranded wire
DC Voltmeter
Earth

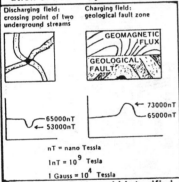

Discharging field: crossing point of two underground streams

Charging field: geological fault zone

GEOMAGNETIC FLUX
GEOLOGICAL FAULT

← 73000nT
← 65000nT
← 65000nT
← 53000nT

nT = nano Tessla
$1nT = 10^9$ Tesla
$1 Gauss = 10^4$ Tessla

Our physical world is typified by ongoing transformations which occur within a discrete range of stress. Beyond the reach of our senses, however, there are others, each with their own landscape, living beings and...ELECTRONS. When we exert stress upon part of a conductor, two things happen. Electrons dislodge from the atomic matrix and migrate in a spiral trajectory toward an unstressed zone. This is inducted voltage. Alien electrons resident in the higher stress level, are then lured by the naked nucleus to assume the orbits vacated by their material counterparts. These in turn begin to flow toward lower stress, and a self-replenishing cycle is begun. When

referenced to 'earth, we now possess a single wire system that requires no actual compensation of charge at its point of origin.

HERE'S THE RECIPE: Place mylar between three layers of copper sheeting. This insulation should extend about 6 inches on all sides. Connect the two outer electrodes to a high frequency Tesla coil. Solder a stranded wire to the centre one. You should now measure an apparent DC voltage between this and electrical ground. THE SECRET INGREDIENT is how to match the self-resonance of your plate capacitor with that of the Tesla coil. When you do, the unit will easily run an electrostatic motor or ignite fluorescent tubes. The generator itself consumes negligible power, since this strategy depends on maintaining a voltage gradient without arc-over.

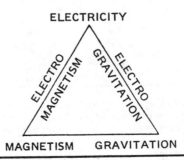

ELECTRICITY

ELECTRO MAGNETISM
ELECTRO GRAVITATION

MAGNETISM   GRAVITATION

I'D LIKE TO DO SOMETHING REALLY SIGNIFICANT WITH MY LIFE .... MAYBE BECOME A GREAT WORLD LEADER AND SAVE THE EARTH FROM NUCLEAR DEVASTATION, FEED ALL THE HUNGRY, OR MAYBE EVEN TRAVEL TO FAR AWAY PLANETS .... BUT WHO AM I KIDDING? I'M JUST A TINY BLACK FLY SITTING ON SOME BALD GUY'S HEAD.

ATToE.

By the way, there are far more efficient shapes for concentrating stresswaves upon a 'neutral centre.' Due to the developmental stage of this project, full disclosure is impractical at this time. Maggie's readers stay tuned to future issues...or send $10 for a generous sampling of illustrated concepts from other sources.

It looks mysterious to the women too, but be assured that the man in this Goma Shobo comic book has no intention of building a bomb. Making a high-temperature superconductor in his hotel suite will keep him busy until dawn.

1949

Burns

King of the Rocket Men, a twelve-chapter Republic serial starring Tristram Coffin, is released. It introduces the now-famous "flying suit": a bullet-shaped helmet, a leather jacket and a twin-rocket backpack (with chest-mounted controls that read: "On, Off, Up, Down, Slow and Fast"). The rocket is propelled by nuclear-powered "sonic propulsion." It is followed by *Radar Men from the Moon* and *Zombies of the Stratosphere* (as well as a television series, "Commando Cody, Sky Marshal of the Universe")

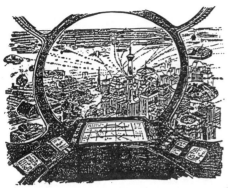

Entering LA, year 2088.

Downtown LA, year 2088.

Traffic Lanes, year 2088.

A Flier's Perspective.

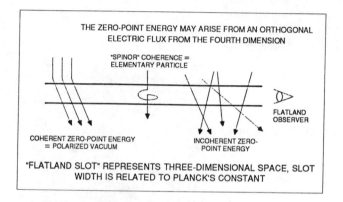

THE ZERO-POINT ENERGY MAY ARISE FROM AN ORTHOGONAL
ELECTRIC FLUX FROM THE FOURTH DIMENSION

"SPINOR" COHERENCE =
ELEMENTARY PARTICLE

FLATLAND
OBSERVER

COHERENT ZERO-POINT ENERGY
= POLARIZED VACUUM

INCOHERENT ZERO-
POINT ENERGY

"FLATLAND SLOT" REPRESENTS THREE-DIMENSIONAL SPACE, SLOT
WIDTH IS RELATED TO PLANCK'S CONSTANT

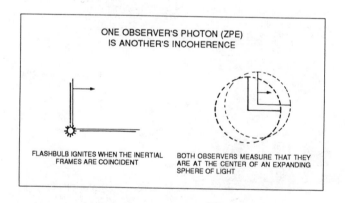

ONE OBSERVER'S PHOTON (ZPE)
IS ANOTHER'S INCOHERENCE

FLASHBULB IGNITES WHEN THE INERTIAL
FRAMES ARE COINCIDENT

BOTH OBSERVERS MEASURE THAT THEY
ARE AT THE CENTER OF AN EXPANDING
SPHERE OF LIGHT

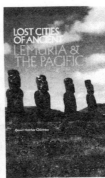

The Lost Cities Series
DAVID HATCHER CHILDRESS

"Explore for lost treasure, stone monuments forgotten in jungles . . . and hair-raising terrain from the safety of your armchair" (*Bookwatch*). "There is a disarming casualness and a kind of festive, footloosed fancy to his tales that are utterly beguiling" (*San Francisco Chronicle*). Rogue adventurer and maverick archaeologist **David Hatcher Childress** takes the reader on unforgettable journeys in search of lost cities and ancient mysteries. Join him as he crosses deserts, mountains, and jungles in search of legendary cities, vast gold treasure, jungle pyramids, ancient seafarers, living dinosaurs, and the solutions to the fantastic mysteries of the past. These books are international bestsellers, translated into both Spanish and Portuguese.

400 pages, 6x9
Travel/Adventure/
Archaeology
Illustrated with photograph
maps, and drawings
Paper

Lost Cities of North and Central America
ISBN 0-932813-09-7, $14.95

Lost Cities and Ancient Mysteries of Africa and Arabia
ISBN 0-932813-06-2, $14.95

Lost Cities of Ancient Lemuria and the Pacific
ISBN 0-932813-04-0, $14.95

Lost Cities of China, Central Asia, and India
ISBN 0-932813-07-0, $14.95

Lost Cities and Ancient Mysteries of South America
ISBN 0-932813-02-X, $14.95

The Lost Science Series
EDITED BY DAVID HATCHER CHILDRESS

The Anti-Gravity Handbook
Revised, expanded edition
230 pages, 7x10
ISBN 0-932813-20-8, $14.95/paper

Anti-Gravity and The World Grid
250 pages, 7x10
ISBN 0-932813-10-0, $12.95/paper

Anti-Gravity and The Unified Field
250 pages, 7x10
ISBN 0-932813-10-0, $14.95/paper

Men and Gods in Mongolia
The Third Book in Our Mystic Traveller Series
HENNING HASLUND

A rare and unusual travel book that takes us into the virtually unknown world of Mongolia, a country that only now, after seventy years, is finally opening up to the West.

358 pages, 6x9
Travel/Adventure/Eastern
Mysticism
Illustrated, 57 photos,
illustrations, maps, all B&
ISBN 0-932813-15-1
$15.95/paper

The Fantastic Inventions of Nikola Tesla
NIKOLA TESLA; edited by DAVID HATCHER CHILDRESS

Nikola Tesla was one of the world's great modern visionaries—and now, at last, there is this comprehensive compilation of the most incredible inventions the world has known. Nikola Tesla held over 1,000 patents for amazing devices, and *The Fantastic Inventions of Nikola Tesla* details the products of this fertile mind with over 200 photos and original patent drawings.

350 pages, 6x9
Science
180 B&W illustrations, 30
B&W photos
Glossary, bibliography
ISBN 0-932813-19-4
$15.95/paper

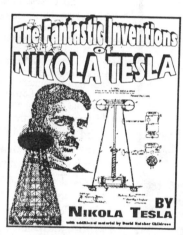
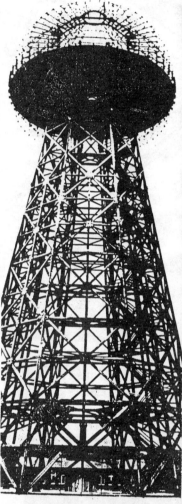

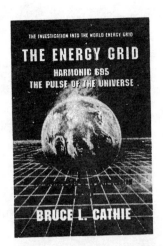

THE ENERGY GRID
Harmonic 695 The Pulse of the Universe
Captain Bruce Cathie.

This is a compilation (updated and corrected) of Air New Zealand Captain Bruce Cathie's first three books. They are all rare, supagesressed and out of print. Exploring the energy grid in great detail there are chapters on unified equations, the mysterious aerials, Pythagoras & the Grid, Nikola Tesla, nuclear detonation & the grid, maps of the ancients, an Australian Stonehenge, & much more. Cathie explores the nature of Light Harmonics and how light hitting the earth creates an energy grid around the earth that is the very basis of physical reality. Cathie's books are a must for the serious student of Anti-Gravity and lost science.

255 pages. 6x9 tradepaper. Illustrated with photographs & diagrams. $13.95. **code: TEG**

THE BRIDGE TO INFINITY
Harmonic 371244
Captain Bruce Cathie

This book is Air New Zealand Captain Bruce Cathie's fourth on the controversial subject of Light Harmonics and the fabric of the universe. Cathie has popularized the concept that the earth is criss-crossed by an electromagnetic grid system that can be used for anti-gravity, free energy, levitation and more. The book includes a new analysis of the harmonic nature of reality, acoustic levitation, pyramid power, harmonic receiver towers, UFO propulsion. It concludes that today's scientists have at their command a fantastic store of knowledge with which to advance the welfare of the human race.

200 pages, 6x9 tradepaper, illustrated with photographs & diagrams. $11.95.
code: BTF

FREE ENERGY SCIENCE

THE MANUAL OF FREE ENERGY DEVICES
Don Kelly.

Combining Vol. One and Two, Kelly describes the viability and progress of each free energy device from Nikola Tesla to the present. Also mentioned are various spin-off inventions as a result of free energy research. Included are chapters on Joseph Newman, "N" Field Machines, Victor Schauberger, John Searle, Nikola Tesla, T. Townsend Brown, Ecklin, Hans Coler and the Coler converter, an early German design allegedly used to power German saucers built during WWII, Wilhelm Reich, Rudolf Steiner, and more. In the closing era of fossil fuels, the fighting of Oil Wars in the Middle East, and the powerful Nuclear Energy lobby promoting the most expensive electricity that money can buy, a book on "free" energy devices belongs in every home.

123 pages, 9x11 tradepaper, illustrated with 100s of rare photographs, patents & diagrams. $14.95. **code: FED**

THE BRIDGE TO INFINITY

HARMONIC 371244

THE ORGONE ACCUMULATOR HANDBOOK
James De Meo, Ph.D.
Forward by Eva Reich

In 1957 all books dealing with orgone were banned and burnt by the U.S. government and its inventor was jailed and assassinated. But today, you can build and experiment with an orgone accumulator of your own. This book presents construction plans and ideas on how to use this simple device for collecting orgone energy and helping to heal plants and animals. In addition, there are suggestions for protecting yourself against toxic energy.

155 pages, 6x9 tradepaper, with diagrams, maps & photographs. $12.95. **code: OAH**

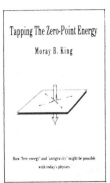

Tapping The Zero-Point Energy
Moray B. King

How 'free energy' and 'antigravity' might be possible with today's physics

TAPPING THE ZERO POINT ENERGY
Free Energy & Anti-Gravity in Today's Physics
Moray B. King.

The author, a well-known researcher, explains how free energy and anti-gravity are possible with today's physics. The theories of the zero point energy show there are tremendous fluctuations of electrical field energy imbedded within the fabric of space. This book shows how in the 1930s inventor T. Henry Moray could produce a fifty kilowatt "free energy" machine; how an electrified vortex plasma creates anti-gravity; how the Pons/Fleischmann "cold fusion" experiment could produce tremendous heat without fusion; and how certain experiments might produce a gravitational anomaly.

170 pages, 6x9 tradepaper, 60 diagrams and drawings. $9.95. **code: TAP**

YHWH
A Book On Ancient Electricity
By Jerry Ziegler

This book is a scholarly and thorough look at the Old Testament and the use of electricity and other high tech devices in ancient times. In 21 chapters, Ziegler has made an impressive study of the Bible; The Ark of the Covenant; Water and Electricity; High Tech devices in ancient Egypt, Early Greek Philosophers and this high technology, the "Dioskouri"; "Archeon"; the "Hymn to Belit"; plus more. 161 pages 6 x9 tradepaper , illustrated with references & b ibliography $9.95 **code: YHWH**

PULSE OF THE PLANET NO.4
Edited by James DeMeo

Seven articles plus features on Orgone Energy and related subjects, including: The Bioelectrical Experiments of Wilhelm Reich translated from the German by Barbara Koopman, M.D.; Why is Reich never mentioned; The Jailing of a Great Scientist in the USA; Cloudbusting in Israel in 1991; more.

190 pages, 10 x12, tradepaper. Illustrated with index, References & Bibliography $19.95 **code: PUP4**

THE VINDICATOR SCROLLS
by Stan Deyo

Stan Deyo's second continues with more technology, conspiracy and Biblical interpretation. In this lavishly illustrated book (mostly in color) Deyo dives into such diverse topics as Atlantis in the Arabian Peninsula; the Ark of the Covenant; UFO technology and other suppressed inventions; unusual material on Nikola Tesla and the secret use of his system of towers; Free Energy research; Government Cover-ups; and more. Only a limited stock has been imported from Australia. Don't miss Stan's videos below.

254 pages, 6x10 tradepaper. Profusely illustrated in color. $19.95. **Code: TVS**

THE COSMIC CONSPIRACY
by Stan Deyo

This amazing book is back in print in a new revised edition with a 40 page new supplement in the back. The Cosmic Conspiracy is a classic in UFO and suppressed technology literature. It was the number one best selling book at one point in Australia, yet most American researchers have never heard of it! Deyo's book is well illustrated and contains chapters on top secret flying saucer research, weather warfare, electrogravitic propulsion systems, underground installations in Australia, suppressed works by Nikola Tesla, theories on Biblical prophesy and the "Illuminati." Don't miss this book! Only a limited stock has been imported from Australia.

272 pages, 6X10 tradepaper, with supplement. Illustrated, $19.95 **Code: TCC**

NEW VIDEO!
STAN DEYO IN CONFERENCE

The second Stan Deyo video, and the sequel to our video **HISTORY OF FREE ENERGY, ANTI-GRAVITY & GOVERNMENT SUPPRESSION.** Deyo is a good speaker, and this second tape is a discussion of various topics, including UFO and government cover-up theories, Biblical Archaeology, Big Brother, the Mark of the Beast, Earth Changes, Prophecy and Prediction. 2 Hours. $29.95 **NEW VIDEO! Code: SDCV**

NEW BOOKS

GENESET TARGET EARTH

David Wood & Ian Campbell

In *GENISIS - The First Book of Revelations*, David Wood exposed a geometric pattern of immense proportions in the district of the Languedoc in the south of France. In this second lavish volume, Wood & Cambell take us on a fascinating trip through the mysteries of Rennes-le-Chateau; Sacred Geometry; Reincarnation, the Cathars and King Solomon's Treasure; the mysterious secret society known as the 'Priory of Sion'; the secret codes of Jules Verne; Time Travel & the Montauk Project; the mysteries of Mars; and much more. Stonehenge, Atlantis, and the World Grid all lead up to the conclusion of a coming catastrophe! In stock. 315 pages, 7x10 hardback. Illustrated with color diagrams, photos and maps. Bibliography. $39.95. **code: GTE**

SEEKERS OF THE LINEAR VISION

Including the Science of Ley Hunting

Paul Screeton and Donald Cyr

The latest large format book from Stonehenge Viewpoint. Covered in this book are such topics as the history of Ley Hunting; Spirituality and the Power Grid theory; Stone Circle Power; Sacred Geometry; Megalithic Alignments; A New Megalithic Key in Scotland; Alfred Watkins and the Old Straight Track, Tony Wedd and the UFO connection; John Michell's "View Over Atlantis"; 25 chapters in all. In stock. 128 pages, 10x12 tradepaper. Illustrated with photos, maps & drawings. Index. $12.95. **code: SLV**

CASEBOOK ON ALTERNATIVE 3

UFOs, Secret Societies and World Control

Jim Keith

Conspiracy and UFO expert, Jim Keith (editor of The Gemstone File) has assembled a powerful and fascinating book into the shadowy world UFO secrets, missing scientists, Cold War collusion between the U.S. and the Soviets, and weird bases on the Moon and Mars. In 1975 an unusual television documentary called ALTERNATIVE 003 was aired on British television which was later followed by a paperback book. The documentary and book detailed the astonishing story of how top scientists around the world were being abducted to provide personnel for secret underground UFO manufacturing bases; how the rich elite of the world have initiated a secret space program of space migration; how secret joint USA-Russian programs have established bases on the Moon and Mars; how UFOs are actually secret government aircraft manufactured in various underground factories on Earth and elsewhere; info on Nazi Flying Saucers; and more. Whether you have the original paperback book or not, don't miss this amazing book! 21 chapters in all. 160 pages. 6x9 tradepaper. Bibliography. $12.95. **code: CA3**

SPACE ALIENS FROM THE PENTAGON

Flying Saucers Are Man-Made Electrical Machines

William R. Lyne

THE HARMONIC CONQUEST OF SPACE

Bruce L. Cathie

The long awaited new book by Capt. Bruce Cathie in his popular *Harmonic Series*., Bruce Cathie's research into the harmonic geometry of the Earth and its energy systems has been going on since he was a New Zealand pilot in the 1960s. His new book recaps and updates all his work to date, and takes the reader further into the realms which connect science with the unexplained. Chapters i this book include: mathematics of the World Grid; the Harmonics of Hiroshima and Nagasaki; Harmonic †transmission and Receiving; the Link Between Human Brain-Waves, the Cavity Resonance between the Earth, the ionosphere and Gravity; Edgar Cayce-the Harmonics of the subconscious; Stonehenge; the Harmonics of the Moon; the Pyramids of Mars; Nikola Tesla the Electric Car; The Robert Adams Pulsed Electric Motor Generator; Harmonic Clues to the Unified Fields, and more. Also included in the book are tables showing the harmonic relationship between the Earth's magnetic field, the 'speed of light,' and anti-gravity/gravity acceleration at different points on the Earth's surface. 260 pages, 6x9. Tradepaper. Illustrated with diagrams and tables. Bibliography. $16.95. **code: HCS August Release**

SPACE ALIENS FROM THE PENTAGON

Flying Saucers Are Man-Made Electrical Machines

William R. Lyne

This large format, tradepaper book goes deep into what the author calls the the "Big Lie" about UFOs. Lyne has produced an excellant and unique book which delves deep into such subjects as Nazi flying saucers and the New Mexico connection; anti-gravity; free-energy devices; Men-in-Black; oil company suppression of technology; and the "space aliens" hoax. A credible book, with some interesting newspaper clippings, photos of Nazi UFO components, and plenty of detailed technical diagrams. Don't miss this book! Well illustrated with photos, charts and technical diagrams. 244 pages. 8x11. Tradepaper. Appendix. $24.95. **code: SAP**

IT'S A CONSPIRACY!
The Shocking Truth About America's Favorite Conspiracy Theories
by The National Insecurity Council

A great rundown of over 60 conspiracy theories. Included are some important chapters on World War II conspiracies, including the SS and the CIA, Rheinhard Gellen, Operation Paperclip, how the Allies created a false Cold War and arms race against the Soviets, Bush's early CIA years, and more. Discussed in detail and objectively (with references and bibliography) are the assassinations of JFK, RFK, MLK, Malcolm X, Oswald and others. Nazi-CIA-banking connections, banking conspiracies, media cover-ups, Saddam Hussein and the Gulf War, the story behind Vietnam, the Red Menace, and more shocking information.

252 pages, 6x9 tradepaper. $9.95. **Code: IAC**

ANCIENT METROLOGY
By Donald Lenzen

This is a fantastic book on Atlantis, ancient maps, the Great Pyramid and ancient measures! Lenzen has crafted a visually stunning and intellectually stimulating book that focuses on measurements, maps, weights and evidence for a highly developed civilization in man's remote past. The cover blurb of this of this large-format hardback says, "Five thousand years ago the human race was crawling out of the Stone Age, yet Ancient Man was using a sophisticated system of weights and measures based on the exact circumference of our Planet. A feat modern man did not achieve until recent times." Don't miss this entertaining book. 104 pages. 8x11 hardback. Illustrated. $14.95. **Code: AMET**

EXTRATERRESTRIAL FRIENDS & FOES
George C. Andrews

Andrews, author of *Extra-Terrestrials Among Us,* maintains in his new book that it is imperative to distinguish between the various occupants of UFOs—human and extraterrestrial. Andrews examines the current popular topics in UFOlogy such as the political implications of a cover-up; alleged manipulation of intelligence agencies; an examination of abductions; an examination of the reported characteristic types and their conflicting motivations and agendas. Other interesting chapters are on Wilhelm Reich, genetically engineered "aliens," and an interesting contribution, "insectoid alien life-forms," from the founder of Borderland Sciences, Riley Crabb. Also with Linda Moulton Howe, J. F. Gille, and more.

342 pages, 6 x9 paperback. illustrated. Index. $14.95 **Code: ETF**

THE MCDANIEL REPORT

On the Failure of Executive, Congressional and Scientific Responsibility in Investigating Possible Evidence of Artifical Structures On the Surface of Mars, and in Setting Mission Priorities for NASA's Mars Observer Spacecraft
Stanely V. McDaniel

This large format, tradepaper book is an analysis of the Cydonia Martian images and a condemnation of NASA and its failure to seriously look into the possible evidence for artificial structures on Mars. Originally prepared as a briefing on the "Martian Hypothesis" for government officials, this scientific document, well illustrated with maps, charts and photos, belongs in every "extraterrestrial archaeology" library. 172 pages, 8x11 tradepaper. Illustrated with photos & diagrams. Appendix. $20.00. **code: TMR**

ELECTROGRAVITICS SYSTEMS
Reports on a New Propulsion Methodology

Thomas Valone, M.A., P.E.

ELECTROGRAVITICS SYSTEMS REPORTS ON A NEW PROPULSION METHODOLOGY

With a Foreward by Elizabeth Rauscher, Ph.D.
Edited by Thomas Valone

An anthology of two unearthed reports of the secret work of T. Townsend Brown shortly after WWII with diagrams and explanations by well-known free-energy and anti-gravity researcher Tom Valone. The first report, Electrogravitic Systems was classified until recently, and the second report, The Gravitics Situation, is a fascinating update on Brown's anti-gravity experiments in the early 50s. Also included Dr. Paul La Violette's research paper on the B-2 as a modern-day version of an eletrogravitic aircraft—a literal U.S. Anti-Gravity Squadron! 116 pages, 6x9 tradepaper, illustrated with patents, graphs & diagrams. $15.00. **code: EGS**

FIELD EFFECT
...The Pi Phase of Physics & The Unified Field
Leigh Richmond-Donahue

Leigh Richmond's brief book is an analysis of the structure of the electron which brings to light the anatomy of black holes, superstrings, and the techniques of evolution. Leigh shows how the "Field" is complete around the universe and all living things. Our galaxy, ourselves, and the electrons of which we are built, function on logical pattern and operative system. Once you find that pattern, the rest falls into place, including free energy, anti-gravity and human enlightenment. 77 pages, 6x9 tradepaper. Illustrated with photos & diagrams. $11.95. **code: FEF**

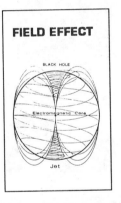

FIELD EFFECT

BLACK HOLE

Electromagnetic Core

Jet

ETHER TECHNOLOGY

A Rational Approach to Gravity Control
Rho Sigma

This classic book on Anti-Gravity & Free Energy is back in print and back in stock. Written by a well-known American scientist under the pseudonym of "Rho Sigma," this book delves into International efforts at gravity control and discoid craft propulsion. Before the Quantum Field, their was "Ether." This small, but informative books has chapters on John Seale and "Searle discs;" T. Townsend Brown and his work on Anti-Gravity; Ether-Vortex-Turbines; and more. Includes a forward by former NASA astronaut Edgar Mitchell. Don't miss this classic book! 108 pages, 6x9 tradepaper, illustrated with photos & diagrams. $9.95. **code: ETT**

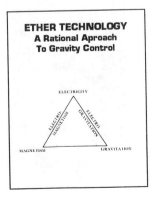

ETHER TECHNOLOGY
A Rational Aproach To Gravity Control

ELECTRICITY

ELECTRO MAGNETISM

ELECTRO GRAVITATION

MAGNETISM   GRAVITATION

ANTI-GRAVITY:
THE DREAM MADE REALITY
The Story of John R.R. Searl
John Thomas Jr.

An important, large format book on the amazing anti-gravity work of British scientist John Searl. Seven chapters tell the story of Searl's invention, his troubles with the British authorities that lead to his arrest and imprisonment, and the road back to starting over. This book contains rare photos and diagrams of Searl's flying disc; plus good technical information on Searl's "Law of Squares," the SEG electric generator; his anti-gravity theories; more. 110 pages, 8x11 spiral bound. Illustrated with photos, schematics & diagrams. $25.00. **code: AGD**

ANTI-GRAVITY
The Dream Made Reality

The Story of John Searl

NEW BOOKS

THE MODERN ALCHEMIST

A Guide To Personal Transformation

Richard & Iona Miller

A first, experiential guide to the process which medieval alchemists represented as the transformation of "lead into gold," or lower substances into higher ones. To Richard and Iona Miller, that transformation goes farther than that—it is an inner change which leads to wholeness, integration, and flowering of the total personality. Alchemy and the language of depth psychology are used in natural transformative process. 223 pages, 6x9 tradepaper. Illustrated. Bibliography & Index. $14.95. code: **TMA**

MYSTERIOUS AUSTRALIA

Rex Gilroy

The first and foremost book on unexplained phenomena down-under is now available. Rex Gilroy is Australia's Charles Fort, and his decades of research are reflected in this impressive book. Subjects include: UFOs and UFO bases; Yowies, Australia's Bigfoot; Giant Skeletons; Pre-Aboriginal Civilizations; Giant Lizards and other reptiles; Underground Mysteries; Little People; Mysterious Mountain Lions and Panthers; Loch Ness Monsters Down-Under; the Tasmanian Tiger; and more. 250 pages, 6x9. Tradepaper. Illustrated with photos, maps and drawings. Bibliography. $16.95. **code: MYA August Release**

TZOLKIN

Visionary Perspectives and Calendar Studies

John Major Jenkins

TZOLKIN is a visionary journey into the heart of an ancient oracle. Jenkins, in both academic and mystical viewpoints, explore the sacred Calendar of Mezoamerica is than a calendar, but an ancient cosmology that includes a mytho-evolutionary system which describes the spiritual and physical unfolding of the Earth. At the heart of this sophisticated philosophy is the Sacred Tree—the Axis Mundi. Prepare for a journey... prepare to to enter the mysterious borderland between night and day! 346 pages. 6x9 tradepaper. Profusely illustrated. $13.95. code: **TZO**

THE SECOND COMING OF SCIENCE

An Intimate Report on the New Science

Brian O'Leary

From the Ashram of India's Satya Sai Baba to the Princeton University laboratories of Dr. Robert Jahn, Ex-astronaut O'Leary provides firsthand accounts of materializations, controlled experiments in psychokinesis, the mysterious crop circles, and other enigmatic occurrences. A co-founder of the International Association for New Science, he argues for an expanded scientific framework that can encompass this challenging spectrum of phenomena and lead us to new understands that may solve our most urgent problems. 168 pages. 6x9 tradepaper. Illustrated with photographs. $12.95. code: **SCS**

NOTHING IN THIS BOOK IS TRUE, BUT THAT'S EXACTLY HOW THINGS ARE

The Esoteric Meaning of the Monuments On Mars

Robert Frissell

Using a mix of fact and speculation, the authors look at the monuments on Mars, the possible presence of extraterrestrials on Earth, crop circles, and our current ecological crisis. Frissell believes that alien beings have co-opted humans to help them change their evolutionary destiny and that these same aliens built the Face on Mars as a transdimensional transmission device to take them to other times and locations and dimensions. Other aliens use consciousness techniques such as meditational breathing, also known as "rebirthing," to travel between dimensions and star systems. 212 pages, 6x9 tradepaper. Illustrated. $12.95. code: **NIT**

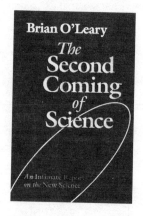

NOTHING IN THIS BOOK IS TRUE, BUT THAT'S EXACTLY HOW THINGS ARE

The Esoteric Meaning of the Monuments On Mars
Robert Frissell

Using a mix of fact and speculation, the authors look at the monuments on Mars, the possible presence of extraterrestrials on Earth, crop circles, and our current ecological crisis. Frissell believes that alien beings have co-opted humans to help them change their evolutionary destiny and that these same aliens built the Face on Mars as a transdimensional transmission device to take them to other times and locations and dimensions. Other aliens use consciousness techniques such as meditational breathing, also known as "rebirthing," to travel between dimensions and star systems. 212 pages, 6x9 tradepaper. Illustrated. $12.95. code: **NIT**

Waiting for the MARTIAN EXPRESS
Cosmic Visitors, Earth Warriors, Luminous Dreams

RICHARD GROSSINGER

METAL POWER
The Soul Life of the Planets
by Alison Davidson

It has been known by ancients, alchemists, and magicians that metals are a powerful force influencing our lives. There are deep mysteries within the metals, hidden connections linking the planets and metals to the Earth's physical and spiritual evolution. Beyond alchemy, metallurgy or astrology, this book will transform your universe with a penetrating view of the magnificent metals and their compelling relationship to the planets.
99 pages, 6x9 tradepaper, illustrated, bibliography, $8.95. code: **MPR**

WAITING FOR THE MARTIAN EXPRESS
Cosmic Visitors, Earth Warriors, Luminous Dreams

Richard Grossinger.

A balanced look at the New Age: shamans, telepathy, prophecy, crystals, pyramids on Mars, extraterrestrials, time warps, vision quests and more.
169 pages, 6x9 tradepaper. $9.95
code:**WME**

EXTRA-TERRESTRIAL ARCHAEOLOGY

THE MONUMENTS OF MARS
Richard Hoagland.

This is an updated and revised book of the original that explores the incredible photos of pyramids, the mile-long face, and a gigantic wall in the Cydonia region of Mars. Careful analysis of the photos has disclosed other monuments and structures, including what is possibly an underground city.
432 pages, 7x10 paperback, photos & illustrations. $16.95 code: **MOM**

THE MARTIAN ENIGMAS
A Closer Look
Mark J. Carlotto

This large format book is a state-of-the-art digital image processing of the controversial Viking photos of Mars and the many structures that are found on that planet. A literal coffee-table photo book of pyramids, faces, roads and unusual structures on Mars. Packed with illustrations and photographs.
123 pages, 10x8 tradepaper, Illustrated $16.95 code: **MAR**

PLANETARY MYSTERIES
Richard Grossinger

An anthology of material, beginning with the pyramids, mile long face on Mars, a gigantic wall and other amazing structures. Other material explores the fractal mysteries of the Mayan Calendar by Jose Arguelles, aboriginal shamans and more.
272 pages, 6x9 tradepaper, photos & illustrations. $12.95 code: **PLM**

DARK MATTER, MISSING PLANETS & NEW COMETS
Paradoxes Resolved Origins Illuminated
Tom Van Flandern

In recent years, astronomy's main theory of cosmology, the Big Bang, has been challenged by many, including Halton Arp (Quasars, Redshifts and Controversies) and Eric Learner (*The Big Bang Never Hapagesened*). Astonomer Van Flandern joins these arguments with a new "Meta Model." The book shows new evidence for the asteroid belt as a "planetary break-up," argues that comets may develop in a different manner than previously supagesosed; and gives the latest information about the possible existence of a tenth planet. A scholarly debate on the current issues in astronomy, including the moons of Mars, Neptune's retrograde moon, the nature of Jupiter's Great Red Spot, and more.
428 pages, 6 x9, tradepaper. Illustrated with photos & drawings. Bibliography. $18.95 code: **DARK**

ALIEN BASES ON THE MOON
Fred Steckling

This book is packed with photos of the strange monuments and structures on the moon. Official NASA photos are used as evidence to show that ancient structures are on the moon, including pyramids. Photos reveal rivers and lakes, clouds, and robot vehicles. Steckling analyzed over 10,000 photos to make his startling conclusions This books raises the question: Who is on the moon?
191 pages, 7x10 tradepaper,125 photos in color and b/w. $15.95. code: **ABM**

EXPLORING INNER AND OUTER SPACE
A Scientist's Perspective on Personal and Planetary Transformation
Brian O'Leary, Ph.D.

This former NASA astronaut probes space science, UFOs, Beam Ships, pyramids on Mars, the living earth and pyramids and more.
182 pages, 6x9 tradepaper, photos & illustrations. $12.95 code: **EIO**

---

§ **24 hour credit card orders** § **phone 815 253 6390** § **fax 815 253 6300** §

ROBERT FLUDD

*Hermetic Philosopher and
Surveyor of Two Worlds*
Joscelyn Godwin

Robert Fludd was an Elizabethan alchemist who wrote voluminously on Rosicrucianism and alchemical thought, applying their doctrines to a vast description of man and the universe. Far ahead of his time in some respects, he recognized the universality of truth, whether from Catholic or Protestant sources, from the Hebrew Bible, from Pythagoras, Plato, or Hermes Trismegistus. Fludd had a genius for expressing his philosophy and cosmology in graphic form, and his works were copiously illustrated by some of the best engravers of his day. All of Fludd's important plates are reproduced in this large format book, making it a visual tour through the alchemical thought of the time. 96 pages, 8x11 tradepaper. Illustrated with 124 diagrams and old prints. $14.95. **code: RFL**

THE KAHUNA RELIGION OF HAWAII

David Kaonohiokala Bray & Douglas Low

From generation to generation since ancient times, the Hawaiians passed down the secret knowledge of spiritual power and sacred knowledge. This is a personal account of Kahuna by the well known native Hawaiian priest 'Daddy Bray' and Douglas Low, a college professor who was in training with Bray for 12 years. Includes detailed instructions on the unique spiritual pathway through ancient Polynesian worlds of creation. 66 pages. 6x9 tradepaper. Illustrated. $8.95. **code: KRH**

THE HIDDEN CHRIST

Was Christ Man or God?
Richard Kieninger

An interesting and inspiring book on early Christianity, Sun Worship, the Biblical "Melchizedek" and a much-needed fresh viewpoint on the life of Jesus Christ as man and God, discussing factually and in depth the hidden source of Christ's reported god-like powers that culminated in His rising from the grave. The book covers Jesus' birth and early life, his fascinating world travels, his connections with Essene Science through the Essene community in Palestine, his training by the Mystery Schools, the ministry of Christ, the politics and pathos of his execution, and the glory of His Resurrection. Kieninger wrote his first book under the name Eklal Kueshana. 132 pages, 6x9 paperback. $8.95. **code: THC**

ALCHEMY & RELIGION

THE HIDDEN TEACHINGS OF JESUS

The Political Meaning of the Kingdom of God
Lance de Haven-Smith

In this radical new view, political philosopher deHaven-Smith examines early Christian writings to discover the message of social reform underlying the teachings of Jesus. The books premise is that Jesus sought to dismantle worldly systems of command and status and replace them with a society governed by a spirit of holiness. Also discussed is how Jesus' prophesies are being fulfilled in the modern era. Huge systems of power and prestige have arisen, but these will be dismantled world-wide by a spirit of holiness. This spirit, the author suggests, can bring about the real kingdom of God, the divine order Jesus urged his followers to establish here on earth. 245 pages, 6x9 tradepaper. Notes, Bibliography, Index. $14.95. **code: HTJ**

CHRISTIANITY AND THE NEW AGE RELIGION

A Bridge Toward Mutual Understanding
Dr. L. David Moore

Many of the "New Age" deny the rich heritage and understanding developed by Christianity, while many of the Christian heritage deny the understanding of the New Age. This book is attempt to bridge understanding between the two, and makes an effort to take a religious-philosophy that spans our planet into a religion that spans the Universe. 9 chapters in all. 244 pages, 6x9 tradepaper. $12.95. **code: CNA**

JESUS CHRIST SUN OF GOD

Ancient Cosmology and Early Christian Symbolism
David Fideler

Hellenistic religious and philosophy scholar David Fideler gives us a book of deep scholarly and esoteric insight. Fideler delves deep into the foundation of modern Western civilization showing us how a lost science of ancient philosophers and architects is part of the teachings of Christ. From the sacred geometry of the ancients to the Spiritual Sun, this book investigates such topics as the Meaning of Logos; the Harmony of Apollo, Mathematics in Antiquity; The Gospel of John and Gnostic Tradition; Mithraic Mysteries; Christianity and the Renewal of Time; Christianity before Christ; Harmony, Strife, and the Growth of Consciousness; more. 430 pages, 6x9 tradepaper. Illustrated. $16.00. **code: JCSG**

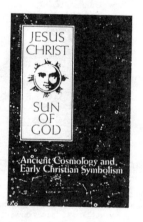

THE MODEL MIND

How the Mind Moves Matter
Peter Bros

The questions of memory, perception and creativity are just a part of *THE MODEL MIND*, Bros explanation of how the mind forms images, how it stores and recalls those images, and why those images, non-material in and of themselves, can drive the body, tormenting it with guilt, destroying it with anger and elevating it to create cities in the sky. Also discussed: the source of neuronic impacts; how does the mind store and recall? 21 chapters in all. 370 pages, 6x9 tradepaper. $16.95. **code: TMM**

ALIEN IDENTITIES
Ancient Insights into Modern UFO Phenomena
Richard L. Thompson

Richard L. Thompson is a Vedic scholar who synthesizes ancient Vedic-Hindu-Dravidian thought into a scholarly look at modern day UFO phenomena. Since 1947 researchers have exhaustively documented the reality of UFOs. But questions still remain about their origins and the intentions of the being who pilot them. In Alien Identities, Dr. Thompson shows that the answers may lie in the records of an ancient civilization with thousands of years of contacts with extraterrestrial races. Startling parallels between modern UFO accounts and the ancient Sanskrit writings of India give fresh insights into the identity and purposes of UFO visitors. Chapters include material on vimanas in Vedic literature; Vedic accounts of Close Encounter phenomena; Machines in ancient and Medieval India; Men in Black; other planes of existence and UFO phenomena; Vedic view of other worlds; more.
512 pages, 6 x9 paperback. Appendix, Bibliography & Index. $19.95. **code: AID**

VEDIC COSMOGRAPHY & ASTRONOMY
Richard L. Thompson

The Vedic literature of India gives a description of the universe that is very difficult for modern people to understand. Topics include the celestial geometry of Bhumandala, mystic powers and higher-dimensional realms, Vedic mathematical astronomy, the dating of Kali-yuga, space travel; the moon flight; astrophysical anomalies; and much more. Includes a fascinating comparison of Vedic versions of the Moon versus the NASA Apollo missions; computer photos of earth's chakra energies at Mount Meru in the Himalayas; the role of Greek influence in Indian astronomy; early Rama Empire trips to the Moon and a clear explanation of the "Vedic World System."
250 pages, 6 x9 paperback. Illustrated with photos, maps, & diagrams. Appendix, Bibliography & Index. $12.95. **code: VCA**

FORBIDDEN ARCHEOLOGY
The Hidden History of the Human Race
Michael A. Cremo and Richard L. Thompson

Richard L. Thompson and Michael Cremo offer a thick (nearly 1000 page) scholarly work that confronts traditional science and archaeology with the overwhelming evidence of advanced and ancient civilizations. Deploying an unexpectedly great number of convincing facts, deeply illuminated with critical analysis, the authors challenge us to rethink our understanding of human origins, identity, and destiny. This book is a step-by-step look at the current evidence for ancient man, and the evidence that is conveniently ignored, and comes up with a scientific attack on the prevailing dogma. Included in this amazing book are chapters or subchapters on anomalous evidence; anomalous human skeletal remains; Living Ape Men?; Chemical & Radiometric testing; evidence for advanced culture in distant ages; more. 960 pages. 7 x10. Hardback. Illustrated with line drawings and photos of artifacts. Appendix, Bibliography & Index. $39.95. **code: FBA**

MECHANISTIC AND NONMECHANISTIC SCIENCE
An Investigation into the Nature of Consciousness and Form
Richard L. Thompson

This book shows that the mechanistic paradigm of modern science cannot account for consciousness and the origin of living species . However, both are tied together in a unified way by the fundamental paradigm of the Bhagavad-gita. The book includes both popular and technical chapters on artificial intelligence; Quantum Mechanics; the illusion of absolute chance; the paradox of Unity and Diversity; Evolution and the enigma of organic structure; more.
254 pages, 6 x9 paperback. Appendix, Bibliography & Index. $12.95. **code: MNS**

Vimana Aircraft of Ancient India & Atlantis

By DAVID HATCHER CHILDRESS

INTRODUCTION BY IVAN T. SANDERSON.
THE COMPLETE VIMAANIKA SHASTRA TEXT.
SECRET LIBRARIES & ANCIENT SCIENCE.
ATLANTEAN AIRCRAFT & TECHNOLOGY.
SANSKRIT SCHOLARS & VIMANA TEXTS.

VIMANA AIRCRAFT OF ANCIENT INDIA & ATLANTIS
David Hatcher Childress

Introduction by Ivan T. Sanderson

Did the ancients have the technology of flight? In this incredible volume on ancient India, authentic Indian texts such as the Ramayana and the Mahabharata, are used to prove that ancient aircraft were in use more than four thousand years ago. Included in this book is the entire Fourth Century BC manuscript Vimaanika Shastra by the ancient author Maharishi Bharadwaaja, translated into English by the Mysore Sanskrit professor G.R. Josyer. Also included are chapters on Atlantean technology, the incredible Rama Empire of India and the devastating wars that destroyed it. Also an entire chapter on mercury vortex propulsion and mercury gyros, the power source described in the ancient Indian texts. Not to be missed by those interested in ancient civilizations or the UFO enigma. 334 pages 6x9 Tradepaper. 104 rare photographs, maps and drawings. $15.95. **code: VAA**

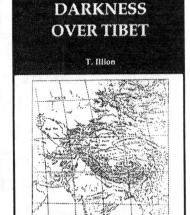

MEN & GODS IN MONGOLIA
by Henning Haslund

First published in 1935 by Kegan Paul of London, this rare and unusual travel book takes us into the virtually unknown world of Mongolia, a country that only now, after seventy years, is finally opening up to the west. Haslund, a Swedish explorer, takes us to the lost city of Karakota in the Gobi desert. We meet the Bodgo Gegen, a God-king in Mongolia similar to the Dalai Lama of Tibet. We meet Dambin Jansang, the dreaded warlord of the "Black Gobi." There is even material in this incredible book on the Hi-mori, an "airhorse" that flies through the air (similar to a Vimana) and carries with it the sacred stone of Chintamani. Aside from the esoteric and mystical material, there is plenty of just plain adventure: caravans across the Gobi desert, kidnapped and held for ransom, initiation into Shamanic societies, warlords, and the violent birth of a new nation.
358 pages, 6x9 paperback, 57 photos, illustrations and maps. $15.95.
code: MGM

VAGABOND GLOBE-TROTTING
Mark Endicott

Ever wanted to take off and see the world but didn't know how to do it? Here is the book that will get you going! This book tells you what to expect, how to plan an extended journey, addresses to write to, organizations that will help you, what you need and safety precautions to take.
200 pages, 6x9 tradepaper, illustrated, $8.95
code: VGT

IN SECRET TIBET
Theodore Illion.

Reprint of a rare 30's travel book. Illion was a German traveller who not only spoke fluent Tibetan, but travelled in disguise through forbidden Tibet when it was off-limits to all outsiders. His incredible adventures make this one of the most exciting travel books ever published. Includes illustrations of Tibetan monks levitating stones by acoustics.
210 pages, 6x9 paperback, illustrated, $15.95. **code: IST**

VIMANA AIRCRAFT OF ANCIENT INDIA & ATLANTIS
David Hatcher Childress
Introduction by Ivan T. Sanderson

Did the ancients have the technology of flight? In this incredible volume on ancient India, authentic Indian texts such as the Ramayana and the Mahabharata, are used to prove that ancient aircraft were in use more than four thousand years ago. Included in this book is the entire Fourth Century BC manuscript Vimaanika Shastra by the ancient author Maharishi Bharadwaaja, translated into English by the Mysore Sanskrit professor G.R. Josyer. Also included are chapters on Atlantean technology, the incredible Rama Empire of India and the devastating wars that destroyed it. Also an entire chapter on mercury vortex propulsion and mercury gyros, the power source described in the ancient Indian texts. Not to be missed by those interested in ancient civilizations or the UFO enigma.
334 pages, 6x9 paperback, 104 rare photographs, maps and drawings. $15.95.
code: VAA

DARKNESS OVER TIBET
Theodore Illion.

In this second reprint of the rare 30's travel books by Illion, the German traveller continues his travels through Tibet and is given the directions to a strange underground city. As the original publisher's remarks said, this is a rare account of an underground city in Tibet by the only Westerner ever to enter it and escape alive!
210 pages, 6x9 paperback, illustrated, $15.95. **code: DOT**

ATLANTIS IN SPAIN
A Study of the Ancient Sun Kingdoms of Spain
E.M. Whishaw

First published by Rider & Co. of London in 1928, this classic book is a study of the megaliths of Spain, ancient writing, cyclopean walls, Sun Worshipping Empires, hydraulic engineering, and sunken cities is now back in print after sixty years. An extremely rare book until this reprint, learn about the Biblical Tartessus; an Atlantean city at Niebla; the Temple of Hercules and the Sun Temple of Seville; Libyans and the Copper Age; more. Profusely illustrated with photos, maps and illustrations.
284 pages, 6 x9, paperback. Epilog with tables of ancient scripts. $15.95. **code: AIS**
•November Publication

IN SECRET MONGOLIA
Henning Haslund

First published by Rider & Co. of London in 1934 as *Tents In Mongolia*, this rare travel book takes us via camel caravan to inner-most Central Asia in search of magic and mystery Mongolia, a country still today largely unknown. Profusely illustrated with photos, maps and illustrations.
284 pages, 6 x9, paperback. Index. $16.95
code: ISM •December Publication

§ 24 hour credit card orders § phone 815 253 6390 § fax 815 253 6300 §

What is *your* idea of excitement?

WORLD EXPLORERS CLUB

Come on our journeys of discovery.

✗ **Diving** to explore the mysterious underwater ruins of Nan Modal, the strange megalithic "crystal city" of the Pacific?

✗ **Paddling** an unexplored Amazon tributary in search of El Dorado, the legendary golden man of South America?

✗ **Hacking** your way through thick jungle searching for lost Inca cities?

✗ **Trekking** across shifting sands in search of the lost city of the Kalahari?

✗ **Climbing** the high Himalayas to visit remote Tibetan monasteries?

✗ **Digging** for buried treasure on remote tropical islands?

✗ **Discovering** previously unknown animal species?

✗ **Sailing** an ocean currents testing new theories of early migration?

If this is excitement to you, then you should be a member of the **World Explorers Club**, a new club founded by some old hands at exploring the remote, exotic, and often quite mysterious nether regions of planet Earth. We're dedicated to the exploration, discovery, understanding, and preservation of the mysteries of man and nature. We go where few have gone before and dare to challenge traditional academic dogma in our effort to learn the truth.

As a member of the **World Explorers Club**, you'll...

✗ Read fascinating first-hand accounts of adventure and exploration in our magazine *World Explorer;*

✗ Have access to the **World Explorers Club** headquarters, its huge archive of history, archaeology, anthropology materials, and map room;

✗ Be eligible to participate in our expeditions;

✗ Receive discounts on **World Explorers Club** merchandise, travel-related services and books (many rare imports) you won't find anywhere else.

Don't miss another exciting issue of *World Explorer*

World Explorers Club • 403 Kemp Street • Kempton, Illinois 60946-0074 USA
24 Hr Credit Card Orders • Telephone: 815-253-6390 • Facsimile: 815-253-6300